THE BUDE CANAL

INLAND WATERWAYS HISTORIES

Edited by Charles Hadfield

The Bude Canal. By Helen Harris and Monica Ellis
The Kennet & Avon Canal. By Kenneth R. Clew
London's Lost Route to Basingstoke. By P. A. L. Vine
London's Lost Route to the Sea. By P. A. L. Vine
The Royal Military Canal. By P. A. L. Vine
The Somersetshire Coal Canal and Railways. By Kenneth R. Clew
The Thames & Severn Canal. By Humphrey Household
The Yorkshire Ouse. By Baron F. Duckham
The Nutbrook Canal. By Peter Stevenson
The Dorset & Somerset Canal. By Kenneth R. Clew
The Leicester Line. By Philip A. Stevens

in preparation

The Ballinamore & Ballyconnell Canal. By Patrick Flanagan
The Birmingham Canal Navigations, Vol. I. By S. R. Broadbridge
The Derby Canal. By Peter Stevenson
The Exeter Canal. By Kenneth R. Clew
The Forth & Clyde Canal. By Graham Matheson and D. Light
The Grand Canal of Ireland. By D. R. Delany
The Grand Junction Canal. By Alan H. Faulkner
The Great Ouse. By D. Summers
The Oxford Canal. By Hugh Compton
The Shropshire Union Canals. By H. Robinson
The Stroudwater Navigation. By M. A. Handford
The Yorkshire Derwent & Pocklington Canal. By Baron F. Duckham

THE BUDE CANAL

by

HELEN HARRIS

and

MONICA ELLIS

With 32 plates and 14 text illustrations
including maps

and a foreword by

David St John Thomas

DAVID & CHARLES : NEWTON ABBOT

ISBN 0 7153 5574 0

Set in 11 pt Garamond, 2 pt leaded
and printed in Great Britain
by Latimer Trend & Company Limited Plymouth
for David & Charles (Publishers) Limited
South Devon House Newton Abbot Devon

Contents

List of Illustrations

PLATES

Photographs not acknowledged above are from the authors' collections

MAPS AND ILLUSTRATIONS IN TEXT

Foreword

THE Bude Canal was surely one of the most extraordinary institutions in rural England. It served one of the poorest and most thinly populated of areas, that bleak plateauland of NW Devon and NE Cornwall; the very idea of there ever having been an extensive canal system there strikes one odd today. And the main cargo was sea sand, rich in lime, needed to tame the sour soil. The curiosities continue: it was partly built with Government money; it was not entirely a canal for the tub-boats were fitted with wheels and were placed on rails on the great inclined planes serving the separate stretches of waterway at different levels. The more you examine the remains and the written records of these planes and their machinery, yet odder it all seems. And miraculously for a time the venture just about paid and there were even spasmodic dividends. Certainly it succeeded handsomely in terms of improving the area's agriculture. To that extent the physical and financial struggles to build the vast engineering work and to keep it and its machinery (especially that of the inclined planes) in working order was well worth while. But what heartaches and what quarrels the managers endured and in what isolation from other grandiose improvement schemes!

I have always had something of a special love affair with the Bude Canal since as a boy I first walked part of its route over a quarter of a century ago. For many years indeed I planned to write the definitive history, always delaying the final execution while collecting more material, paying more visits to decaying wharfs where life has stood still since the last tub-boatful of sand was unloaded, and talking to elderly folk who remembered the inclined planes at work and retold the legends of hope and sorrow that surrounded the canal's rise and decay. Alas, pressures on my time increased as the publishing house producing this work grew rapidly, and for some years any thoughts of writing the detailed history have had to be put aside. And in a sense it was a journey I was happier to contemplate than make since I got almost too close to the subject. That I should be writing the introduction to a work others have written and I am publishing is therefore a fair outcome. It's not quite the book I might have written, but it's almost certainly better for being more dispassionate.

Today, when snatching a few hours to unwind from a fast and furious life, I sometimes still go to see the Bude Canal, sitting on the harbour breakwater marvelling that there were such few accidents to sailing ships bringing their tiny cargoes of coal from South Wales, wandering around the works at the bottom of the great plane with the words the hard-pressed managers wrote in their Day Books still in my mind, and just relaxing at Tamar Lake, the canal's reservoir and now a nature reserve. Apart from Bude itself, the area remains poor and today the railway (vibrating with the famous LSWR 4-4-0 T9 'Greyhounds' when I first knew it) is in as ruinous a state as the canal it superseded. Coal reaches Bude by road, and rarely indeed are the canal's sea lock gates opened—though a proposal to brick them up to save cost some years ago resulted in one of Bude's best ever attended protest meetings. Our little systems truly have their day. Yet they were worthwhile, and though ceased to be in practical terms (apart from a short stretch for pleasure boating) the Bude Canal can still give pleasure . . . through reading its remarkable history so ably set out here, through exploring its remains on the ground with the help of the authors' itineraries.

DAVID ST JOHN THOMAS

CHAPTER I

Preceding History

EVEN today the culm measure land of north-west Devon and the bordering areas of Cornwall poses problems for the men who farm it. This wide tract, bounded on the west from Boscastle to the Taw estuary by the Atlantic coast, and extending eastwards (excluding Dartmoor) in places for up to 50 miles, is composed of soils which are inherently heavy, acid, wet and cold. In summer, when the hedges resound with the wistful song of the yellowhammer and the bubbling cry of the curlew is not uncommon, indigenous rushes, meadow sweet and hemp agrimony flourish in pastures where they can escape the farmer's vigilance, pastures which in winter must be grazed sparingly or not at all if they are to avoid being poached out of all usefulness.

Two centuries ago this stretch of country consisted mostly of old moorland, with the land in a state of poor permanent pasture. Much of it was weed-infested and out of tilth, and gave the general impression of a wild and forbidding waste. Tolerating a standard of living which was little more than a mere subsistence, members of the sparsely spread farming community grazed their lean bovine stock on the so-called grassland and tilled a proportionally small acreage of inferior arable, predominantly to wheat.

But in the last quarter of the eighteenth century even this remote back-country was becoming aware of the ripples of the tide of agricultural improvement currently gaining ground in other parts of England. Observers noticed a change in this far western countryside. Pastures in places were becoming more luxuriant, the uplands were carrying more sheep and corn growing was on the increase. New enclosures were being made out of the moorland, and the deep ditches, excavated for making the traditional hedgebanks, provided an aid to drainage and a run-off for the copious springs of iron-charged water with which the land was inundated. In the reclamation, and for the maintenance of fertility and good tilth, manuring had become of increasing importance.

Established methods of manuring consisted chiefly of the spreading of dung, and the burning of the sod, or beat—a process also termed 'Devonshiring'. The value of lime was also recognised. Certain parts of the area were more advantageously situated for obtaining supplies of burnt limestone for application on the land than were others; some were within easy reach of the kilns which burnt material from Devonian limestone quarries, others, in the north, were served by estuary or coastal kilns which used limestone and culm (an inferior coal-like substance) brought in by sea from South Wales. On the other hand, the localities in the hinterland of Bude, and in the watershed of the upper reaches of the River Tamar were not proximate to such facilities, and here farmers resorted to another procedure, the application of sea sand from the beach at Bude.

The use of sea sand as a manure, practised also elsewhere, was no new idea. Marshall, writing in 1796, describes it in the district as being 'beyond memory, or tradition'. Sand from Widemouth, in Bude Bay, is known to have been fetched regularly by teams of pack-horses and used on the Werrington lands of Tavistock Abbey in the fourteenth century. The method of its use in the cultivation of wheat on the old moorlands of the district is outlined by Charles Vancouver (1808) as follows:

> the first operation is to spade and burn the surface, spread the ashes, and dress with about 60 seams, or 160 bushels of bude sand per acre. Instances near the borders of Cornwall are not unusual, for 100 seams to be applied, but fifty only are provided

by covenant as a dressing for wheat, and the two succeeding crops of oats or barley. This sand is either applied alone or mixed with fresh mould, and ploughed lightly under with the beat ashes into ten-furrow ridges, and then sown with from six to eight pecks of prepared seed wheat per acre, harrowed in, and producing on an average through the Holdsworthy quarter, eighteen bushels of 62lbs each, per acre'.

When used as a top-dressing on grassland Vancouver noticed that the sand never failed to bring forth sweet and valuable herbage in a succession of seasons.

The great value of Bude sand lay in the high proportion of calcium carbonate—approximately 50 per cent—in its composition. In addition to the neutralising and floculating effects which this produced on the acid soils to which it was applied, physical improvement also resulted from the mixing of the loose and large particled sand with the close, clinging clay. Sand lying farthest inshore, blown or washed there because of its lighter weight, was considered the best on account of its superior calcium content. On certain soils the beneficial effects of dressing with sand were found to be more lasting than with burnt limestone, effects which can still be detected on some farms in the Holsworthy district.

As the sand was available free of charge, loading and transport were the chief factors to be reckoned with, so that obviously the cost of using it as a manure increased in proportion to the distance from the sea. Until the latter part of the eighteenth century, roads from the coast being no more than narrow boggy trackways, the sole means of transport was by packhorses which carried the sand in 'potts' or strong coarse panniers supported across their backs. From the 1770s, however, there was some improvement in the conditions and wheeled transport was introduced. Consequently the movement of sand could be achieved more quickly and in greater quantity, but after a few years was causing concern locally due to the wear and tear on the highways. Vancouver wrote:

The prodigious traffic through the parishes of Holdsworthy,

> Pancrassweek, Bridgerule, Lancelles, Marham Church, and Stratton, to Bude-bay in Cornwall, for sea-sand, requires some arrangement, that in its operation might convey relief to those parishes, on account of the heavy road-rates their inhabitants are annually liable to for keeping up their respective parish roads.

But by the time these words were written ideas had already been put forward for a completely different method of transporting the sand, one that seemed almost revolutionary in this remote area but was in line with the movement already established in other parts of the country farther north—the construction of a canal to bring the sand from the sea-coast to the interior.

EARLY PROPOSALS AND THE ACT OF 1774

The first proposal for a Bude Canal was put forward in 1774 by John Edyvean, a Cornishman who at the time was pioneering the construction of the St Columb Canal in his home county. A route for the canal was surveyed by Edmund Leach and John Box, who said that it would be 'of great public utility' when called in, with Edyvean, to give evidence to the House of Lords Committee giving consideration to the Bill.[1] The landlords concerned, and seven of the county's justices of the peace having signified their approbation, 'An Act for making a navigable Cut or Canal from the Port or Harbour of Bude, in the hundred of Stratton, in the County of Cornwall, to the River Tamar, in the parish of Calstoke, in the said county' was passed, receiving the Royal Assent on 24 May 1774 (14 Geo III c 53).

The effect of the proposed canal was to be the linking of the Bristol and English Channels by way of the Tamar's navigable waterway, this early example of the ambitions for cross-country connections between seas perhaps being inspired by the construction of the Trent & Mersey Canal, then in progress. It was intended that not only sand, but also Welsh coal, stone, lime and other manures, mining products, timber, agricultural goods and domestic supplies would all be transported on the system.

Although the direct distance between Bude and Calstock is only 28 miles the canal was to follow an exceedingly tortuous and circuitous route of around 90 miles. The plan was for it to pass from Bude through Marhamchurch, Poundstock, Whitstone, Bridgerule, North Tamerton, Boyton, Werrington, North Petherwin, Tremaine and Egloskerry to Launceston, and from there by way of Laneast, Lewannick, North Hill, Linkinhorne and Stoke Climsland to Calstock. But the vast degree of deviation in its course, made necessary by the features of the countryside and a clause in the act which prohibited cutting more than 39in deep on the canal's lower side, was not at the time considered disadvantageous, since a maximum of parishes and outlying farms would thus be served. Also, the uplands where the canal was to run had the poorer soil, so that the land which would benefit most from the sandy manure was also the cheapest to buy.

The canal was to have a width of 21ft at the top and 12ft at the bottom, with sufficient depth for tub-boats of 10 tons. The land to be used was not to exceed 53ft in width—including the towing-paths, which were to be fenced—except for places where boats were to pass each other or be turned or docked. A clause in the Act also required that near the mansion of Whiteford, in the parish of Stoke Climsland, the canal should be arched over and carried underground for a distance of 300ft. To negotiate the hilly country it was to be necessary to raise the canal a considerable height above the terminal points, and this it was proposed to do by means of five inclined planes, making use of engines of a particular construction apparently invented by Edyvean himself. Trucks were to be used on the planes, necessitating the transfer of cargoes at each. The summit level was to run for a distance of 68 miles, with three of the inclined planes at the Bude end and two taking the canal down to the connection with the Tamar. The first northern plane was to be of 54ft rise, the second—6½ miles farther along—of 120ft rise, and the third—a further 4 miles distant—of 66ft. The summit was, therefore, to be 240ft above sea level. The two planes at the southern end were each to be of 120ft, the first one descending from the end of the summit at a point 2½ miles from the Tamarside terminus where, at Kelly Rock, the fifth plane was to be sited.

Various local people gave their support to the project and the raising of £40,000 capital in £100 shares was authorised, plus a further £20,000 if required. Twenty shares per person was the limit, each share carrying one vote. Navigation on the canal was to be free on payment of the rates set for the commodities, which were low; duties per ton of slate, stone, sand, culm, coals, timber and manure were not to exceed 2s (10p), or 3s (15p) per ton for goods, wares or merchandise. No wharfage was to be charged for goods remaining at the wharves less than twenty-four hours, and only 2d or 3d per ton for that remaining a week. Powers enabled the Commissioners to raise or lower tolls at their discretion, provided the shareholders received their interest at 6 per cent.

Probably on account of the somewhat unattractive financial prospects from the low rates of toll and the limited amount of capital which had been authorised there was no apparent great keenness for the work to commence. By the following year still nothing had been started and there was talk of abandoning the project.

SMEATON'S REPORT: SUGGESTIONS FOR LOCKS AND DAMS

In 1777 the engineer John Smeaton, whose works had included the reconstruction of the Eddystone lighthouse and planning a route for the Forth & Clyde Canal, was consulted on the practicability of the scheme. After viewing the terrain Smeaton had no doubt that the canal was physically practicable in its proposed course and mechanics, but, estimating the cost at £119,201 he was concerned about the economics. Observing that the county of Cornwall, so frequently intersected with valleys, in general seemed ill adapted for the making of canals,

Smeaton, in his report to the promoters dated 8 January 1778, considered various modifications and alternative ideas, estimating the cost of each, and the respective financial returns which could be expected.

One suggestion was for shortening the proposed length by altering the planned route between Bude and Launceston and carrying it over Greena Moor, which, though it would have necessitated an additional lift—and a corresponding drop—would have reduced the length by 15 miles and avoided considerable meanderings in several small valleys. Further considerations were the prospects of constructing the canal in just its northern part, from Bude to Launceston only, either by the line originally planned or by the cut across Greena Moor; but such limitation, though appearing as a safer venture, failed, Smeaton felt, fully to comply with the original transport objects of the scheme. When, however, he observed the valley of the River Tamar, its openness and conveniently gentle fall, Smeaton was struck by another thought—the ease with which the river above Calstock could be adapted as a navigable waterway by means of locks and dams. Accordingly he prepared another scheme. In this it was suggested that six locks should be built in the Bude River to make it navigable for 3½ miles upstream, then 6 miles of canal constructed, rising by two inclined planes to the summit level and descending by one to join the River Tamar in the vicinity of North Tamerton. From here ten or perhaps more locks would be needed in the 15½ miles to Greston bridge, and a possible branch made up the tributary River Kensey to Launceston. From Bude to Greston bridge (the name is now corrupted to Greystone) Smeaton estimated the cost at £46,109. Though he made no costings for continuing the waterway to Calstock Smeaton was satisfied this could be done for less money by locking the river than by other means.[2]

But even after Smeaton's report no decision was apparently taken, and no start made. The Act of 1774 provided that powers obtained under it should cease after ten years, and as nothing was done within this time the scheme was abandoned.

LEACH'S PREFERRED PLAN AND IDEAS FOR INCLINED PLANES

An attempt at revival was, however, made in 1785 when Edmund Leach produced some fresh ideas based on Smeaton's report. These he put forward in a book entitled *A Treatise of Universal Inland Navigation.* Leach was evidently fascinated by the prospects of inclined planes as a means of lifting boats from one level of a canal to another, which in the British Isles had so far only been employed in the construction of the Tyrone Canal in Ireland. These he considered as a less expensive alternative to the locks which Smeaton had proposed and he drew up a new plan, shortening the distance of the route originally designed and introducing planes to effect the necessary lifts.

Leach's preferred plan was to take the route from Bude Haven up the valley along a one-level side canal, with a possible branch to Widemouth, a distance of 8 miles, ascending to the cliff top at Bude and at Widemouth in each case by an inclined plane. From the foot of Week St Mary Hill a plane was to raise the boats and their cargoes up to the 9½ mile 'canal of Partition', which would perambulate the head of the Week St Mary valley to a place near Whitstone called the Pack Saddle and, receiving its supply of water from the Tamar by an aqueduct from Shernick Moor (north of Bridgerule), reach a suitable point for descent into the Tamar valley, near Tamerton bridge, by another plane. For the remaining stretch, downstream to Calstock, Leach again advocated a side canal and a further inclined plane instead of the river navigation and locks. The cost of the full scheme, 40¾ miles from Bude to Calstock, and its ancillaries, longer in distance than that estimated by Smeaton, Leach reckoned as £53,100, a considerable saving on the figure of £88,740 at which his opinion put the original scheme of 1774.

It is interesting to note Leach's design for his 'mechanical hydraulical machine' by which he proposed operation of the inclined planes, adaptable for use either on canals or in mines. Unlike Edyvean's earlier idea of reloading cargoes at each ascent and descent, Leach planned to carry the boats in their entirety. He recommended using oak or other wood for construction of the line of the plane between the two levels, 22ft 6in in breadth (for canals) divided longitudinally into equal parts, with wooden ledges at the sides and along the middle. Allowing for the ledges, the path on each side would be 10ft in width. A strong wall, with sluices, was required at the end of the upper canal section and between this and the head of the plane a horizontal platform of two equal parts was required. Rollers on the canal side of this wall were to facilitate passing the boats in and out and further sluices at the edge of the canal would allow water to be let out for powering a waterwheel. Special vehicles were to transfer the boats and their cargoes up and down the paths. Roughly triangular in section, these were designed to carry the boats horizontally, with a cistern to contain water for use as a counterbalance. Rollers were to be provided between the bottom of the vehicle and the plane and others on which the boat would rest and by which it could easily be transferred to and from the canal. A hole in the bottom of the cistern was to allow water to enter and the vehicle to sink below water level when taking on boats from the canal's lower section; this ballast water would be allowed to discharge as the vehicle ascended the plane, to effect correct balance with its descending counterpart. Operation was therefore partly by counterbalance and partly by power from the waterwheel, which served as an over-all control.

In spite of their ingenuity, however, there were no immediate results from Leach's plans, and still the canal remained unbuilt.

THE NUTTALLS' REPORT

In the early 1790s, at about the time when a new ambitious scheme was being planned for constructing a 30 mile canal from Morwellham up the Tamar valley to Tamerton bridge[3] (later completed in part, as far as Gunnislake, as the Tamar Manure Navigation) there was renewed interest. This was strongly supported by the Holsworthy landowner Lord Stanhope, a man of modern ideas and of scientific and engineering inventions, and by other influential local gentlemen. Early in 1793 the possibility of a canal from Bude through Holsworthy to Hatherleigh was considered and a sum of money subscribed, but a preliminary survey showed that a larger amount would be needed. Later the same year a more detailed study was made by two other engineers, John and George Nuttall, whose report was presented to a meeting held at the White Hart Inn, Holsworthy, on 25 October 1793, with Lord Stanhope in the chair.

In order to negotiate the intervening ridges of hills the Nuttalls ascertained that the canal would have to be carried at a height of 473½ft to cross the easternmost ridge by means of an 11ft cutting at Hammett's Lane, north-east of Holsworthy, and they considered the various available methods of lifting at their disposal. The use of steam engines and perpendicular lifts was one but this was rejected, the hills involved not being of sufficient perpendicular height. Another was the system which employed ropes and whimseys powered by horses or steam, but for this also the hills' slopes were found to be unsuitable. Locks, too, were ruled out for various reasons; firstly due to the potential insufficiency of water for the number of locks required, secondly because the time that would be needed in passing through the locks would not enable enough boats and their cargoes to use the canal each day, and thirdly, if the locks were made big enough to take the maximum cargo in fewer boats, the cost of construction would be too great. Another factor was that locks would deprive numerous mills in the district of their supplies of water.

The method which the Nuttalls favoured was the use of 'iron-rail roads' in conjunction with the canal. They found that it would be possible to construct the canal in four level sections and to connect them by lengths of railway, amounting to almost 4 miles of railway in all. The first length, of about ¼ mile, would be at Bude, rising to the canal's first level section of a little under $2\frac{1}{2}$ miles which would take the route into the Stratton valley, north of Marhamchurch. From here the second length of railway, of about $2\frac{1}{4}$ miles, was planned to run up the Cann Orchard valley and eastwards to the south of Thorn. Near Thorn the second canal level was to start. This, of nearly $8\frac{1}{4}$ miles, would pass under the Launceston–Kilkhampton road (south of Red Post) by a $\frac{1}{4}$ mile tunnel and then, curving northwards, cross the road which runs west from Bridgerule bridge and the one from Holsworthy to Bude west of Tamarstone, and proceed to the bank of the Tamar at Burmsdon. From a point near Thorn a branch of the same level was to run southwards towards Merrifield and continue to Tamerton bridge where, by a double inclined plane, connection could be made with any future canal constructed northwards from Launceston. A side cut of the branch level could if necessary be made from Merrifield to Whitstone and Week St Mary.

From Burmsdon a third length of railway, of slightly over $1\frac{1}{4}$ miles, was to ascend through Pancrasweek to Lana, where a connection with the canal's summit level would be made. This level, the principal section, would be at least 50 miles in length; there was to be a short side branch near its starting point northwards to Wood 'to meet the sand road that passes between Virworthy and Lutton'. The summit level was to run eastwards over Ugworthy Moor, near Vognacott, Burscott and Yellowland and bend south over Manworthy Moor, through the 'Holsworthy Poor Lands' to a field near Hammett's Lane, and over Blagdon Moor by Lamerton and Herdwick. East of Holsworthy the canal would cross the Hatherleigh road and then run southwards, in the vicinity of Simpson, Southcott and Hollacombe to the head of the Claw valley on Dunsland Moor, where it would meet the main road from Okehampton. From a point near Hammett's Lane another branch of the level would proceed north-eastwards to run north of Woodacott, south-west of East Wonford, east of West Wonford, and westwards below Thuborough and over Soldon Moor to Waterland Mill, Bradworthy, where it would come near the River Waldon. The

Waldon was considered the best supply of water for feeding the canal, and the connection here would make it unnecessary to draw supplies from the Tamar.

In addition to the tunnel near Thorn, four others were proposed. One was to be on the canal's lowest reach, near Bude; one south of Holsworthy, east of Chasty; another north of Woodacott, and another south of East Wonford. The tunnel north of Woodacott was to be the longest, of nearly ½ mile, the other three here mentioned each being under ¼ mile.

In order to continue the canal towards Hatherleigh and to circumnavigate the River Torridge it was planned to cross the River Waldon near Waterland and, following the Waldon valley, to carry the canal eastwards, below Sutcombe and near Gratton and Buttermoor in Milton Damerel parish to the top of Knat's Wood. Here the summit level was to end and a drop made by means of a double inclined plane down to the Torridge, which would be crossed just above Newton Mill where the valley is narrow. The fourth reach of the canal, of nearly 14 miles, was to proceed on the level eastwards from here, below Allacott in Shebbear parish to within a mile of Hatherleigh.

The Nuttalls' total estimated cost for the scheme, with its 75 mile waterway, was £32,404.50. They calculated also that more than 250,000 acres would benefit from its facilities, enabling, it was reckoned, the realisation of a 10 per cent annual revenue on the capital sum.[4]

It was Lord Stanhope's plan to use small boats of 2 tons and to carry these on the rail sections suspended between a pair of wheels about 6ft in diameter, which would be drawn up and let down by horses. The Holsworthy meeting considered the Nuttalls' report and resolved unanimously to apply to Parliament for an Act. John and George Nuttall were appointed to complete the plans for the scheme, Lord Stanhope was thanked for his 'indefatigable Zeal and Perseverance' and, while further subscriptions apparently continued to be received, the meeting was adjourned sine die.

ROBERT FULTON: WHEELED BOATS AND WATER-POWERED INCLINED PLANES

At about this time, in the September of 1793, a three-year correspondence began between Robert Fulton, a young artist recently arrived from America, and Lord Stanhope. Fulton seems quite suddenly to have become possessed of an inspiration for the designing and engineering of inclined planes, and although his ideas were not entirely new they pre-suggested methods which were employed at a later date, when the canal was eventually built. The first suggestion which Fulton made was for the working of the inclined planes by the use of water in containers to provide weight for descent, a method which, Lord Stanhope pointed out, had already been put forward by Edmund Leach. On 8 May 1794, however, Fulton obtained a patent for the use of a double inclined plane, still using the water cistern or caisson arrangement, each caisson being carried level on four or eight wheels. He saw the plane itself as working by counterbalance: either the descending load would be heavier than that coming up, or additional water would be added to the descending caisson. He provided, however, for capstan or water-wheel power to get the boats up the reverse slope from the upper pound of the canal and into the caisson: 'each cistern is to have a small carriage on which the boat is to float, and be drawn out of the canal, up the side-plane on the cistern. . . . The power to draw the boat on the cistern, may be obtained by a capstan or a waterwheel'. The wooden boats were to be rectangular in shape, of 2–4ft width; those of 4ft width were recommended to be 20ft long and 2ft 10in deep, giving a capacity of 4 tons. The two pairs of wheels underneath were to be from 6 to 10in diameter, each pair being made in one piece with the axle.[5]

In a book published in 1796, *Treatise on Canal Navigation*, Robert Fulton further expounded his proposals for narrow canals and inclined planes and the use of shallow rectangular boats with wheels, but advocated alternative methods of lifting.

One of these was for using power produced by a water tub descending in a perpendicular shaft to raise loaded boats up the incline, the boats being attached by individual chains to an endless chain passing up one track of the double inclined plane and down the other and around wheels at top and bottom. A variation, recommended for use on planes of smaller ascent, involved the use of a waterwheel sited beside the top of the plane, instead of the descending tub, for providing the lifting and braking power.

These plans of Fulton's are of considerable interest because of their applications in the Bude Canal's actual construction a quarter of a century later. For the bucket-in-the-well method—the principle on which the Hobbacott plane later operated—Fulton advised, for the sake of economy, making the plane between selected hillside points that would provide the greatest possible rise at one time. The slope of the hill, he explained, should be formed into a regular incline between the canal's two water levels, of an angle less than 45 degrees, and on this the two planes, each of 2ft 1in width, made parallel, 6ft apart. On a bedding of rubble, overlaid with either timber or coping stones, the iron rails were to be set at a gauge of 2ft 1in. The bottom of the plane was required to level out to a more gradual slope for a distance of about 60ft to the lower canal, with the track meeting the water, both here and at the top, by a gentle curved bridge, so as to prevent the undersides of the boats scraping the planes. The pit or well, of 10–11ft diameter, was to be made at the top of the plane, of equal depth to the difference between the two levels of canal, with a drainage sough from its bottom connecting with the lower waterway. The bucket, of capacity in excess of the required 10 tons of water, was specified to be of 9ft diameter and 5ft depth, made of wood or sheet iron, with a 12–18in diameter hole in the bottom fitted with a valve device which would open and discharge the water as the tub struck the bottom of the pit. Water from the upper canal was to be conveyed by a trough into the tub, which would be suspended by two or more chains from a drum wheel over the mouth of the well, with balance chains and a weight on the opposite side for returning the emptied tub to the top of the pit. A shaft would convey power from the drum to the upper wheel for the plane's main chain; this wheel, set at the same angle as the plane, was to be of 8ft diameter, so that its extremities would come exactly opposite the centre of the two planes. The corresponding wheel at the bottom of the incline was to be of the same size, but set horizontally. Power from the inclined wheel at the top was also to be used for drawing the boats out of the upper canal. Fulton described carefully the mechanics of the system and the way it would work; timber he considered could be rafted along the waterways, bound together, and at the planes could be floated on to special 'carriages' for ascent or descent. He also gave close attention to time and analysed the various operations, estimating actual passage over the plane taking as little as 1½ minutes.

The use of a waterwheel for providing power was recommended by Fulton where the perpendicular ascent to be achieved was under 200ft, perhaps as little as 20ft. In the book he gave a design for a single plane of the latter height. All the planes of the Bude Canal were of double track, but the principle of the overshot waterwheel, to which the leading chain connected, was the same.

REVIVAL OF INTEREST AND JAMES GREEN'S PLAN

Doubtless due to the ensuing war with France and other preoccupations plans for the canal were pushed into abeyance for several years. Then, probably in 1814, the prospects for reviving the scheme were discussed by a Mr Harward of Tackbeare, Bridgerule, and a neighbour, a Mr Braddon of Newacott. It was felt that the work of construction would be a means of employment for the great numbers of men returning from the fighting, and as Braddon had in his possession Nuttalls' report of 1793 it was agreed that Harward should go to London, taking the report with him, and consult there with Lord Stanhope, and see if he would give the project his renewed support. Lord Stanhope was fully in agreement and it was decided to arrange a meeting in London with the other interested people during the following winter. But Napoleon's return from Elba to France and the renewal of the war prevented this taking place, and in 1816 Lord Stanhope died. Messrs Harward and Braddon did not let the matter drop, however, and, obtaining the local support of Mr George Call and others, they approached the succeeding Lord Stanhope who readily added his approval to the plan.[6] Consequently, in 1817 the engineer James Green and surveyor Thomas Shearm were called in and invited to survey a line.

Green's instructions were to direct his attentions to the opening of a communication between Bude and certain points in the interior, the arrangement of the line being left to his discretion, and to survey the haven of Bude and report how far it was capable of improvement. The points in the interior at first specified were Thornbury, Holsworthy and as far eastwards as time allowed, and Tamerton bridge. During the work the committee of subscribers decided to extend the survey and Green was asked to continue his inspection from Tamerton bridge to Launceston. He was required to divide his plan and the estimates into several parts so that there would be no difficulties if it was decided not to proceed with the whole extent of the work.

Green presented his report to the subscribers on 14 April 1818. Two alternatives were open to him, the construction of either a canal or a railway. A railway, he considered, would be the less suitable, on account of the rise of ground and nature of the soil, and the frequent changing of horses that would be entailed. The usual type of canal, with locks, he also saw as impracticable, for the same reasons as did the Nuttalls in 1793. Green's recommendation, therefore, was for a canal on a small and inexpensive scale, 19ft wide at the top, 10ft at the bottom, with 3ft depth of water. He proposed boats of 5 tons each, four

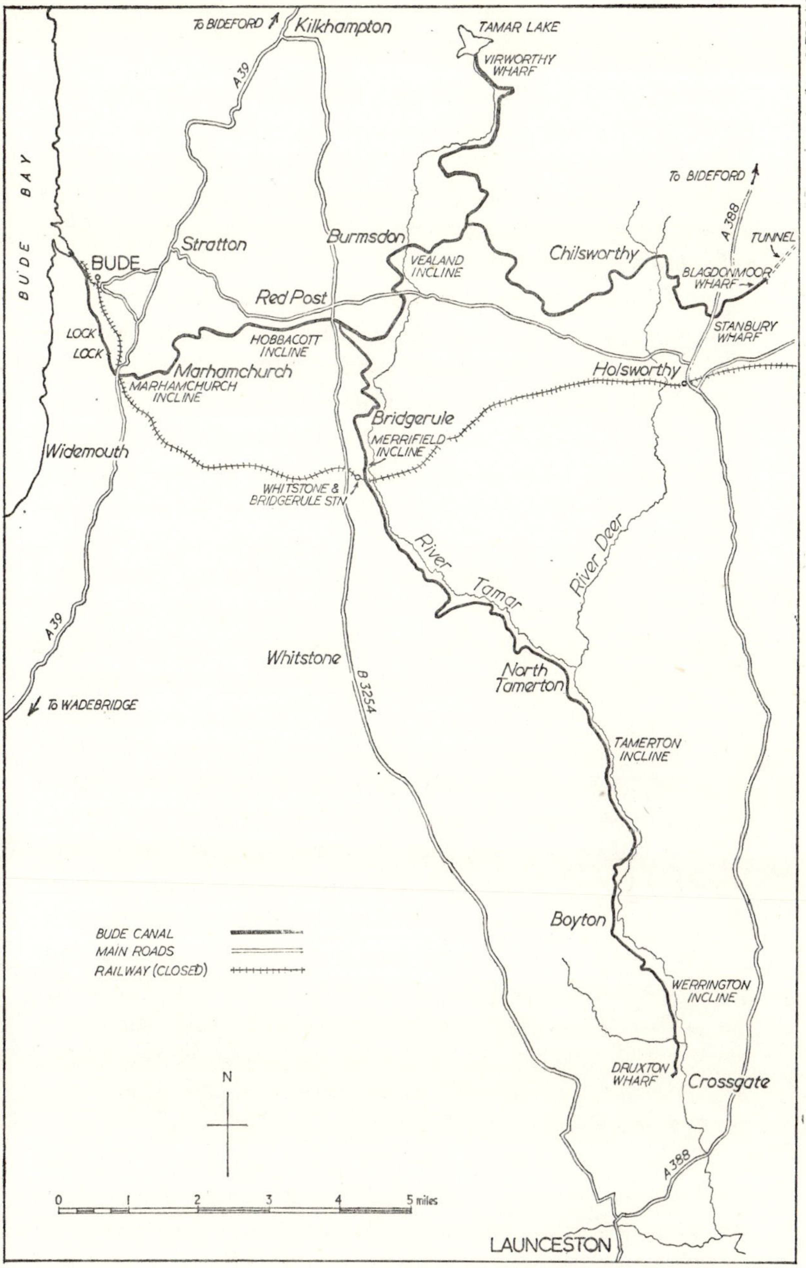

FIGURE 1. The Bude Canal as built

being drawn by one horse, and to overcome the ascent by means of simple water-powered inclined planes, up which the boats would pass. The countryside was in his opinion particularly favourable for such planes, which he estimated could be constructed at a third of the expense of locks, worked with one-third the amount of water and save more than four-fifths of the time of passage.

With regard to the harbour, where, even then, some considerable trade was carried on, Green observed that the chief drawbacks were its smallness and the difficulty of entry. In his view some protection from the strong seas and deepening of the harbour were needed, and would lead to the still much greater trade which could be expected from the extensive 'back country'. His recommendations were that a breakwater, 10ft above spring tide level, should be constructed to close the gap between the mainland of the south side of the mouth and Chapel Rocks, and that the channel of the river should be deepened to a point near the sandhills, where the water could be ponded back and released periodically in a body to provide scouring power for keeping the outer channel clear of accumulated sand.

Although he had also surveyed and estimated for a smaller length of canal, unconnected with the harbour and on the eastern side of the river, Green's actual proposal was to commence the canal at the bend of the river with a lock, capable of admitting vessels of 40–50 tons, and to continue the line along the west side of the valley towards Helebridge, a distance of almost 2 miles. The channel here was to be 24ft wide at the bottom, 37ft 6in at the surface, and 4ft 6in deep. The rise of 25ft Green planned to accomplish by means of two further locks, taking three united streams from near Helebridge as a supply of water. Barges of 40 tons, which had taken up sand at low water, entering the first lock at flood tide, could thus navigate the canal to the foot of the first inclined plane, where the cargo could be transferred to the small boats. The locks could also be used by four or six of the small boats at one time when necessary, so that outgoing goods could be delivered directly to vessels lying in the harbour by small boat and barge. A further, possible later, improvement to the scheme, Green here remarked, would be the construction of a ship lock at the entrance, capable of admitting vessels of 70–100 tons into the basin, on the banks of which warehouses might be built.

Inland from Helebridge, from which it was proposed to rise by inclined planes at Marhamchurch and Thurlibeer, Green's projected line for the canal was basically similar as far as Burmsdon and North Tamerton to that suggested by the Nuttalls, but in its extremities it differed. At Burmsdon he planned an aqueduct to cross the Tamar and an inclined plane near Veala to take the line to the summit level. This, the main trunk of the canal, requiring cuttings and valley crossings, Green proposed making as short as possible, favouring side branches rather than an over-circuitous route. One such branch he proposed to Virworthy, partly for making supplies of sand conveniently available to the parishes of Bradworthy, Sutcombe and Milton Damerel, and also to serve as a feeder for the canal's water supply, for which he planned construction of a 60 acre reservoir on the River Tamar at Alfardisworthy.

From Holsworthy another branch was proposed to run towards Thornbury, but, though he estimated for it, Green did not advocate continuing the canal beyond this on the Nuttalls' suggested line as he considered the expense unwarrantable. Eastwards from Holsworthy the canal was to continue via Dunsland Moor, Graddon or Chilla and Northlew and, crossing the Okehampton–Hatherleigh turnpike at Croft, to Brightley bridge, near Okehampton, with tunnels at Windmill Cross (near Holsworthy), Dunsland Moor, Graddon, Northlew and Croft. Though he terminated his survey at this point, Green was confident that the line could be carried farther in the direction of South Zeal, North Tawton, Bow and Crediton from which possible future communication could be made to Exeter, Bideford, Barnstaple, South Molton and Tiverton.

Green planned to have one inclined plane descending from

the summit level on the line to Tamerton bridge. Though time did not permit his complying with the belated request for information on the possible extension to Launceston in the form of a full survey, he was of the opinion that the country was suitable, but recommended termination short of the town, at Rushgrove Mill, to avoid crossing the river in Werrington Park. The extension, he felt, might also have the advantage of bringing about completion of the Tamar Manure Navigation, thus opening a communication between the seas to north and south.

The total cost of the construction of the canal and the improvements to the harbour Green estimated at £128,341, and he foresaw an annual return on capital of at least 10 per cent from sand alone. In addition to the much-needed sea sand it was expected that coal and other imports from Wales and the Bristol Channel would be carried to the interior and that, in addition to timber and agricultural products, the proposed extremities of the waterway would enable granite and manganese (from mines between Launceston and Tavistock) to pass outwards. The extension to Okehampton would also bring into reach limestone quarries needing coal for their kilns and from which lime could be carried.[7]

In spite of the inevitable gloomy pessimists who maintained that 'the farmer will continue to bring home his sand by the old conveyance' the Committee was convinced that the canal would be worth while. Returns from parishes on the line indicated that 28,038 tons of sand were taken up annually, and with the cheaper form of carriage which the canal would provide it was predicted that even greater quantities would be used and the capital more than supported. In the words of the Committee's report to the subscribers, presented at a meeting held at Launceston on 15 April 1818:

> By the present conveyance, the farmer, distant ten miles, can only go three times a week during the spring and summer months, and, with three horses, draws only eight seams, or one ton, that distance in one day. By the improved plan, one

> horse will draw twenty tons the same distance, so that three horses will bring sixty tons, and, going at all seasons of the year, will, if employed every alternate day for the purpose, bring one hundred and twenty times as much as can be procured at present by the same number of horses.[8]

A further meeting of subscribers was held at Launceston on 5 November when measures were discussed and the extent of the canal determined.[9] Early the next year, following proposals to extend still farther south beyond Launceston, Lord Stanhope approached the Duke of Bedford on the subject of repealing the Tamar Manure Navigation Act (the Bude subscribers having purchased the Tamar company's shares) and carrying the extension on the Cornish instead of the Devon bank.[10] This prospect, which appealed to the duke since it would have avoided his estate at Endsleigh, did not, however, materialise, when, in 1819, the Act of Parliament was obtained.

CHAPTER 2

Under Construction

THE COMPANY

In 1819 the Bude Harbour & Canal Company was formed. Of the 330 shareholders, leading subscribers were Earl and Countess Stanhope, Sir William and Lady Call, George Call (the company's first chairman), Sir Arscott Molesworth, Sir Thomas Dyke Acland and William Arundel Harris. Members of the Coham family of Black Torrington, and of the Kingdon family—including the vicar of Bridgerule—were also subscribers, as was James Green, the appointed engineer, who invested a substantial sum. Many names on the list were those of Exeter business men, the most prominent being John Blackmore, a merchant of 143 Fore Street; unlike the local landowners, most of the Exeter shareholders can have had little territorial interest in the proposed canal and subscribed their money, presumably, simply in what seemed to them a sound financial investment.

It was probably because of contacts in Exeter, as well as reputation and experience he had gained in that locality, that James Green, an energetic man of far-seeing mind, usually self-confident and at times opinionated, was appointed engineer. Born in Birmingham in 1781, the son of an engineer, Green worked with his father until 1801, after which he was employed under Rennie on various engineering projects throughout the country. In 1808, after submitting successful plans for the new Fenny Bridges in east Devon (to replace the previous Fenny Bridges, disastrously collapsed only eighteen months after rebuilding) Green was appointed Bridge Surveyor

Page 33 *Stucley* of Padstow and tub-boats at Bude

Page 34 (*above*) The approach to Bude; (*below*) The lower wharf at Bude when the 4ft-gauge sand railway was in use, showing two of the tip-up trolleys and the tub-boats ready to be filled with sand

for the county of Devon, and in 1818 was made Surveyor of Bridges and Buildings for the county, a post which he held until 1841. At that time it was customary for county surveyors also to undertake private work, and after settling at Exeter Green advertised his availability. In 1810 he produced for Lord Rolle a plan for a canal from Bideford to Torrington in north Devon, which, when it was later built under Green's supervision in 1824–5, carried tub-boats and included a water-powered inclined plane. Probably Green's first actual constructional canal work after his arrival in Devon was that of engineer to the Exeter & Crediton Canal, which was started in 1810 but shortly abandoned after excavation of only half a mile. The Bude Canal appointment was probably his first important assignment. It was one, however, which took only part of his time: from 1820–7 he was employed in enlarging and extending the Exeter Canal, he produced a report for the proposed Looe & Liskeard Canal and later one for the Chard, and in 1822 reported on the possibilities of a tub-boat canal with inclined planes to link the English and Bristol Channels through Devon and Somerset. It is probably in connection with the eventual Grand Western Canal that Green is best known; his plan for seven vertical lifts and an inclined plane on an extension, begun in 1832 to link Rennie's earlier section with Taunton, was adopted but not entirely successful.[1]

Undoubtedly because of its convenience alike for Green and for numerous other members, and particularly, perhaps, because of the interests of the Aclands who owned much land at Bude but lived near Exeter, the Bude Harbour & Canal Company was based at Exeter, the 'metropolis of the south-west', and most of its meetings were held there, usually, in the early days, at the New London Inn in the city.

THE ACT

Very shortly after the formation of the company, on 14 June 1819, an Act was obtained 'for improving the Harbour of Bude, in the county of Cornwall, and for making and maintaining a navigable Canal from the said Harbour of Bude, to or near the village of Thornbury, in the county of Devon and divers branches therefrom, all in the said counties of Cornwall and Devon' (59 Geo III c 55). Power was given to raise £95,000 in 1,900 shares of £50 each, and an additional sum of £20,000 if needed, either amongst themselves or on a mortgage of the undertaking.

There had necessarily been some pruning of Green's original plan, resulting in reduction of the estimated cost by nearly a third: £91,617 was the final figure arrived at which included £4,618 for the harbour works. The idea for the canal's eastward extension from Holsworthy across Dunsland Moor and towards Okehampton was scrapped, and the terminus of the southern branch which at first was to be at Rushgrove Mill, Launceston, was, on the Bill's third reading, brought back by about 3 miles to Druxton. Bason Lane, between Thornbury and Bradford, was still to be the waterway's north-eastern terminating point

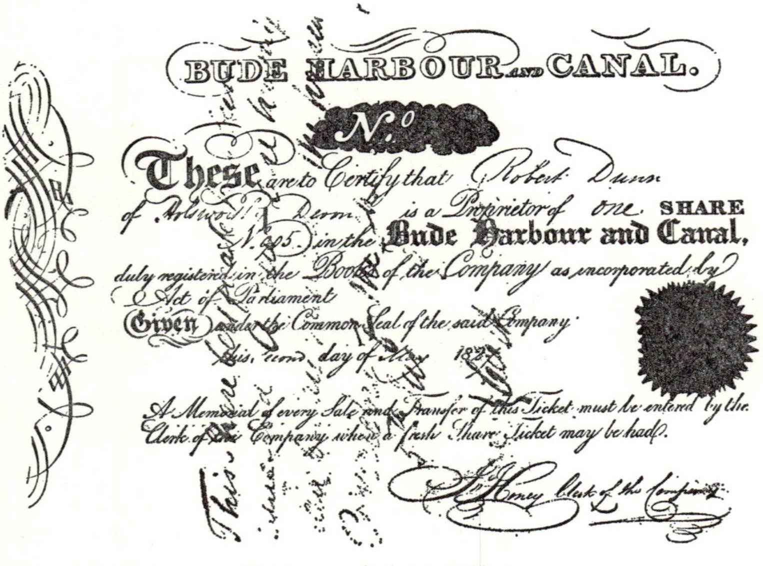
BUDE HARBOUR AND CANAL.

No.

These are to Certify that Robert Dunn of [illegible] Devon is a Proprietor of one SHARE N. [illegible] in the Bude Harbour and Canal, duly registered in the Books of the Company as incorporated by Act of Parliament

Given under the Common Seal of the said Company this [illegible] day of [illegible] 18[illegible]

A Memorial of every Sale and Transfer of this Ticket must be entered by the Clerk of the Company when a fresh Share Ticket may be had.

[illegible] Clerk of the Company

FIGURE 2. A share certificate

and provision was made for a collateral canal or cut from near Veala through Virworthy to Langford Moor, which would serve both for navigation and as a feeder from the reservoir which was there to be constructed. Another navigable cut, of 1½ miles, was authorised to run northwards from Burmsdon to Moreton Mill. The length from Bude to Hele, Marhamchurch, was to be made suitable for barges and the remainder of the system for tub-boats, the total length of the proposed canal amounting to just under 46 miles.

INAUGURATION OF THE WORK

No time was lost in starting the work of construction which officially began on 23 July 1819. A report in a subsequent issue of the *Exeter Flying Post*[2] gave a graphic account of the official inauguration, which took place with much ceremony and merrymaking. At the request of the subscribers Lord Stanhope made the long coach journey from his seat in Kent specially for the occasion and when he arrived, on the day prior to the commencement of the work on the canal, was given a great welcome by the people of Stratton and its surrounding neighbourhood. While a band played the people unharnessed the horses from his carriage and themselves drew him into the little town, amid shouts of applause. Handsome entertainment was provided at Stratton which was 'numerously and respectably attended' with George Call chairing the proceedings. Amongst the fare on the festive tables were some 'prodigiously fine fat venison' which had been sent by Lord Rolle; there were speeches from Mr Call and Lord Stanhope and the Bude Harbour and Canal was well and enthusiastically toasted.

On the following morning, a Wednesday, the ceremonies started early. At 6 o'clock the bells were rung and a *feu de joie* was discharged by a group of old volunteers. Later a large crowd of an estimated 12,000 people gathered on Bude's Summerleaze Down. Under James Green's direction a crane and engine had been placed at the head of the intended breakwater and spades and wheelbarrows at the place for the canal's commencement. Presently the firing of a small cannon heralded the approach of a 'grand procession of ladies and gentlemen', headed by Lord Stanhope and Lord Mahon and preceded by flags and bands. The stately parade halted at the appointed place and formed a circle around Lord Stanhope who then laid the breakwater's first stone and made an eloquent speech, applauded with cheers 'echoing and re-echoing among the rocks and cliffs'. Then the procession continued to the site of the canal basin where Lord Stanhope cut the first sod, again greeted by much applause, after which the bands played 'God save the King' and 'Rule Britannia'. Everyone then returned to the down

> where, to commemorate these great events, there were erected booths, and also an immense and magnificent temporary structure covered with canvas, and ornamented with evergreens and festoons of flowers. In these booths banquets were provided, at which the viands (although the best of their kinds) were less gratifying to the animal appetite than the sensibility excited was to the liberal mind. In the large structure, which from the quantities of beautiful flowers, might properly be called 'the Temple of Flora' after the company had been regaled, dancing commenced in two long sets, which continued with highest glee, gaiety and hilarity for several hours. On other parts of the down various gymnastic exercises and feats of activity were performed—such as wrestling, running, etc etc. Ten hogsheads of cider and many thousand cakes were given to the populace, and proper refreshments were provided for everyone.

CONSTRUCTION: PROBLEMS AND PROGRESS

After the frivolities of the inauguration celebrations the hard steady slog of construction work began and the hilly countryside between Bude and Holsworthy, over which the line of the canal had been marked out, became a scene of much activity. One can imagine the talk that must have gone on in the villages and farmhouse kitchens of the neighbourhood where the canal work would have undoubtedly provided an unending topic of conversation. Though they were to benefit from more easily available sand the farmers, with their innate caution and reluctance for change, would understandably have found plenty to grumble about, especially when faced with the excavations actually cutting across their land and the inconveniences frequently caused. Local tradespeople probably welcomed the influx of navvies, their numbers running into hundreds, who came with their alien dialects to work on the scheme, many of them most likely veterans of the recent wars, others probably migrants from completed canal work elsewhere.

Some idea of the way in which work on the canal progressed can be formed from the day-to-day entries in the journal of the canal company's inspector, which is preserved at the Castle, Bude. From the journal, which covers a period of fifteen months from 19 October 1820 to 11 January 1822 it is evident that throughout this time work was going on simultaneously on different parts of the route, from Bude to a little beyond Holsworthy, on the reservoir and feeder, and along the 'Lanson' (Launceston) line from Red Post to Tamerton bridge, as well as on the harbour, breakwater and locks at Bude. The inspector, in whose hand the majority of the entries are written and whose signature is also recorded, appears to have been a man called John Panchen, who was also one of the original subscribers mentioned in the Act of 1819. Another, unsigned handwriting also occurs in places, while during the last two and a half months of the spell covered the duties were performed by one John Kingdon. The inspector's work inevitably entailed much walking, some of which on occasions was in the company of the chairman, Mr Call, who lived at Vacy, North Tamerton, and whose signature also appears in the journal.

The first four days of the journal's record are typical and fairly representative of the general order:

Thursday 19 October 1820 In company with the Chairman walking to the Tunnel near Hammetts Lane and inspected the whole of the line of the Canal from that

Thursday 19 October 1820	point back to Roydon Moor. Some few of the fences out of repair, but in general the whole was in good order. No workmen on this part of the line.
Friday 20 October	In company with the Chairman examined the whole of the Line from Little Bridge to Red Post . . . then took the Eastern line from the branch Bridge to Pancrasweeke Valley . . . then back to Burmsdon and up the navigable feeder to the Reservoir on which branch found about sixty men at work. Inspected and found the Culvert in good condition.
Saturday 21 October	. . . examined line from Red Post Bridge to Bude, observed some leaks in the canal at Thurlibeer.
Sunday 22 October	Remained at Bude in company with the Chairman watching the effect of High Tide and Seas on the Breakwater and Sea Locks, counted on the works 340 men.

Mention of the tunnel, which re-occurs in a number of subsequent entries, is of interest, and something of a mystery. It obviously refers to the commencement of the line from Holsworthy to Thornbury, which was, in fact, never completed. Amongst other instances, under 4 November 1820 there appears: '. . . another party (of men) at the eastern part of the Tunnel . . .', for 18 November 'crossed the Tamar at Kingford Mill and examined the Holsworthy Line from Veala over Pancrasweeke, Chilsworthy, Manworthy embankments to East side of the Tunnel. . . . Parties working in Pancrasweeke valley and the Tunnel', on 26 December '. . . Deep cutting each end of the Tunnel but not finished. Canal cut from west end of Tunnel say one mile or within 300yds of the road near Stanbury Cross' and, for 27 January 1821 'Crossed the Tamar at Burmsdon and inspected the Holsworthy Line to the Tunnel. The only men working are at the latter place where are 17, and 9 raising clay for making bricks at Arscott'. Although this section of the line was not proceeded with, and there are now no remains of the tunnel, it is possible to trace the commencement of the canal's projected continuation from cuttings which are visible beyond Blagdonmoor wharf. Obviously a certain amount of labour was expended here, but to no avail.

Much of the inspector's work was involved with looking into complaints of various sorts and in dealing with poles and hurdles used for fencing, over which there seemed to be continual trouble. The grievances were mostly raised as the result of damage caused by some act or omission of the constructors. We read such items as 'Mr Brumell complains of damage been done cutting through a field of oats at Vealand which he estimates at £2 10s and requests may be remunerated', 'Rogers of Rodd's Bridge complains that Mr Green's men and carts had broken Gate Post and claim remuneration', 'Piper of Burmsdon reported Horse fallen into deep cutting near his house and killed, claim by him £1' and 'for want of rails an ox just lost this week coming over one of the Bridges with a load of manure'. Some farmers evidently became irate, for one entry 'Saw Richard Lyle concerning his field of clover which was damaged some time ago by Mr Green's carts passing over it' is followed by others 'Measured Mr Lyle's field the damage done by carts as near as I am able to judge is about $\frac{1}{6}$ of an acre of Clover' and 'Mr Lyle of Cross makes complaints of not being paid for the damage done in his field and threatens to summon the Person who drove the cart across it before some Justice of the Peace'.

Naturally, as a new section of the canal came under construction it was important for barriers to be erected immediately against the adjacent farm land, or livestock would stray. This was not always done with sufficient care, and brought further farmers' complaints about animals trespassing. There was trouble, too, from pillaging of the fencing materials and, it appears, by the throwing down of temporary fences by boatmen passing with materials along the lower reaches. A constant check had to be kept by the inspector resulting in such entries as 'Took account of the Hurdles and Poles belonging to the Company amounting to 333 of the former and 305 of the latter', 'Hurdles much wanted at Virworthy and Dunison. Hired a cart to go to Bude for 20 Hurdles to be placed at Virworthy where wanted at 6 shillings for Cart and Man and 1 shilling for fencing' and 'No of Hurdles 707 Poles 335 and Posts 229 on the whole of the line'.

There are no records of the total number of men engaged on the canal's construction, although the various journal entries give an indication of the deployment and how the work progressed. For 23 October 1820: 'One gang of about 20 men working on the Incline Plane at Hobbacott Down', 26 October: 'York's gang of 21 men working at the Embankment and Stockey's gang of 24 men on Virworthy Valley with about 22 men at the Reservoir', 10 November: 'Parties working on Obbacott (*Hobbacott*) Downs House on that part of the work in progress. The culvert at the Foot of the Great Incline finished . . .', 14 November: '. . . The cutting from the Reservoir to Puckland advancing fast . . . Bridge at Puckland in progress . . .', 20 December: 'Large gangs at work at Shernick, Anderton and some on Thurlibeer. Shafts building up and Machinery getting ready to clear them of water.' Entries in 1821 include: 16 January: 'Inspected line from Hobbacott to Bude Breakwater and counted 190 on this section', 22 January: 'Took the Lanston (*Launceston*) line from Little Bridge to Red Post and from that point to Bude, 14 men working along Thurlibeer and Mead Park, some Puddling the Bank and others throwing up the Fences, some at Cann Orchard and some at Marham Church, some at the Inclines and some on Whalesboro Meadow where the course of the stream is turned into the Canal. Many men employed boating down stuff for raising the towing path below, and Masons on the waste weir at Whalesboro and some on the lock at Rodd's Bridge, another party on Sir T. D. Acland's works and some in the Breakwater . . .', 8 March:

'... There are upwards of 60 men at work filling up the Dam Head at the Reservoir' and 14 September: '... Two gangs of men at Lana Moor on to Pancrasweeke Valley, one 18 the other 28. A gang of 24 men at Thorn Moor, Parnacott Valley 28 men, Hoggs Park 18 men. At Manworthy Valley 15.' We also learn that the aqueduct across the Tamar at Burmsdon was commenced on 11 May that year.

Work reports on several of the canal's important constructional features form part of the journal's illuminating commentary. There was evidently much activity on 27 December 1820 at Hobbacott as the works for the plane took shape: 'One of the whims ... at work and one shaft [*for the well and bucket system*] expected to be clear of water this night. Plane partly forward and each shaft sunk about 70ft and the adit driven from the foot of the Plane to Cann Orchard, the ground open but no part can be said to be completed.' Whims (horse-powered hoists for raising waste material) were at work here day and night. Progress was often far from easy; a few days earlier, on 14 December, the inspector wrote: 'The shaft at Marhamchurch Incline for the Water Wheel to work in is fallen in and a gang clearing it.' The necessary river alterations below here were being done around the same time; on 29 November there was 'A party of men cutting through a field near Hele Bridge to turn the course of a small stream that now runs parallel to the road' and subsequent entries record that the river's old course was being filled up. A little later, on 21 April 1821, there was a trial run from the basin through the sea lock and on to the beach by one of the barges which, at the time, it was intended should be used for bringing the sand from beach to basin. 'Took the Barge no 1 out the Sea Lock and put on board her about 24 tons of sand pm at tide time got her into the Basin, the Barge drawing 3ft 6in aft and 2ft 10in forward. Laid down the Buoys with mooring stones to place Barge on the sand. The Easternmost Buoy 100yds from the narrowest part of the Channel.' In fact this method was not the one later employed, as will shortly be seen.

FINANCIAL DIFFICULTIES—THE CANAL COMPLETED

By 1821, although work was proceeding apace, doubts were being felt over the canal's financial prospects. In a letter to Sir Thomas Acland, one of the principal shareholders and owner of land at Bude, dated 6 May 1821, John Carew after a visit to Bude wrote: 'The concern appears to be going on prosperously and by no means warrants the depression which the shares have experienced. Your Basin is complete and filled with water; and a handsome thing it is. Whether it will immediately pay its way is not so clear, and many persons who have watched the progress of the canal do not expect that you will reap those advantages which Green pointed out.'[3] In the same month the committee decided to suspend construction of the line to Thornbury and a few months later, on 5 January 1822, the general assembly determined that the tolls should be mortgaged.

This was contrary to the convictions of Lord Stanhope who subsequently set out his observations in a letter to the Bude Harbour & Canal Company, dated 27 April 1822, for consideration at the general assembly. Confident of the favourable prospects for the sale of sand on the length of the canal shortly to be opened, the earl, who accepted reluctantly the necessity for suspending the Thornbury line, strongly recommended that construction of the extension to Druxton should also be postponed until new shares could be created to provide the necessary funds, and he urgently warned against the mortgaging of tolls which he considered would be an act of suicide by the company. He put forward a suggestion for transporting sand from beach to canal by a railroad instead of by the barges, a scheme estimated to incur the additional expense of £600 but which, besides being more convenient, would enable sand to be delivered 1d per ton cheaper and the capital required for providing barges to be reduced. Lord Stanhope also expressed his view that trade on the canal should be carried on by individuals, as had been originally intended, rather than by a private co-partnership as was now proposed. He differed from the chairman, George Call, in his more optimistic estimates of the expected consumption of sand and the income to be gained from it and he advocated imposing low tolls to bring high sales and large profits. He wrote: 'It is the object of the consumers to purchase a large quantity; it is the object of the Company to sell a large quantity; neither object can be accomplished unless the price is low.'

Lord Stanhope's indication of not seeing eye-to-eye with Mr Call, and a statement that he was 'aware an attempt had been made to excite disunion among the Proprietors' show that all was not harmonious within the company and an exchange of letters early in 1823 are evidence of further differences which led to George Call ceasing to be chairman. It appears that, early in 1822, the canal company's committee of management adopted the suggestion of the appointment of a sub-committee, consisting of George Call, the Rev Davis and a Mr Vowler, to conduct its affairs 'on the ground' where the canal was being cut. This was sanctioned by the general assembly at its meeting in May that year and the sum of £300 per annum agreed on as payment for expenses—£100 to each. The chairman had already been involved in continual work on the company's behalf for which in his position he felt responsible, which often necessitated overnight lodging at Bude. Many difficulties had arisen concerning the purchase of land for the canal, the cost of which had generally been calculated too low, owners were clamouring for settlement and there were constant demands for compensation for damage and trespass—all matters which were borne out by the journal entries. This took considerable time to deal with and as the committee of management could not easily be convened George Call felt fully justified in sharing the sum allotted to him and the two fellow sub-committee members who assisted in judging the claims.

The company's clerk, George Thorn, and various other people, however, saw the matter in a different light and doubts on the legality of the arrangement were expressed in the first of a series of sycophantic public letters which Thorn wrote to Lord Stanhope, dated 15 January 1823. In it the money paid was referred to as 'salary' though in fact this term had not been used. Lord Stanhope gave his considered reply in a letter of 18 February, written after he had sought further legal advice. He pointed out that the resolution had been very loosely worded and the duties of the sub-committee insufficiently defined, but that it seemed to him the three members had been exercising functions which properly should belong to the whole committee, of which five members constituted a quorum. As such the Act of Parliament was being violated. He also condemned the allowance of 'salaries' which were having to be paid not out of income but from capital. George Call's term as chairman ended abruptly, in the same month. He evidently felt deeply the implications of the letters, which in a letter to the company's proprietors, dated 23 September 1823, he described as 'a mass of malicious misrepresentation'.

By this time there had been a dropping off in work on the canal and the main line was nearing completion. On 3 October 1822 the *Exeter Flying Post* recorded: 'Upwards of 200 labourers, called navigators, lately employed on the Bude Canal, have arrived in Taunton for the purpose of commencing their operations on the projected Taunton & Bristol Canal.' And early the next year a meeting was called for 2 April to consider opening the canal for trade, to authorise working of the standby steam engine installed at the Hobbacott plane and to sell, let or employ barges or boats, the property of the company, on the line. By June the canal was declared ready for trade and on 8 July 1823 the official opening of the completed portion—from Bude to Holsworthy, to the reservoir and to Tamerton bridge—took place. Barges and boats belonging to one of the principal shareholders, Mr Blackmore of Exeter, described as 'that respectable and liberal character' (although later, it appears from correspondence, obstinate over the payment of tolls) were loaded with manure (sand) and merchandise for Holsworthy. When they arrived the committee of management (Lord Stanhope was absent on this occasion) 'supported by the neighbouring gentry . . . marched into the town of Holsworthy in procession, the band playing "See the conquering hero comes", and hailed by the acclamation of the populace of the surrounding country. The many peals from St Peter's tower also marked the novel event as the proudest day ever witnessed in the district. . . .' A dinner was provided at the Stanhope Arms, numerous toasts were drunk and there was music and much hilarity.[4] The ensuing prospects were bright; there was an immediate brisk demand for sand, tolls taken in the first three weeks after the opening appeared to exceed the expenses of keeping the canal in service,[5] and there was a dramatic rise in the price of shares.

Capital funds, however, had run low and on 26 June 1823 application was made to the Exchequer Bill Loan Commissioners for a loan of £20,000, which was backed by a report from Thomas Telford, acting as the Commissioners' adviser. The sum applied for was reduced by the company to £16,000 when, in the following year, subject to certain conditions, the loan was sanctioned.[6] Work on the southern line, from Tamerton bridge to Druxton, was commenced in the summer of 1824, by which time there were already 100 boats operating on the canal.[7] A further £4,000 was, however, still needed, and was obtained in another loan from the commissioners before the work was finally completed in 1825. The total cost of constructing the canal (minus its planned but abandoned extension beyond Blagdonmoor, Holsworthy to Thornbury, and the cut to Moreton Mill) and the cost of maintenance and other expenses up to the time of its completion, is believed to have amounted to just under £120,000.

CHAPTER 3

Early Working Days

FEATURES AND ROUTE OF THE NEW CANAL

FROM the entrance to the sea lock on the Bude shore to the terminal points at Blagdonmoor, Holsworthy, and Druxton bridge, Launceston, and with the feeder arm from the reservoir, the canal's total completed length amounted to 35½ miles; substantial iron mileposts were set at appropriate intervals and smaller ones at the interim quarters. The rise from sea level to summit on the Holsworthy line reached a height of 433ft, while on the Launceston line there was a drop from the intermediate 375ft to 205ft.

An important and vital feature of the construction, though not a part of the canal itself, was the breakwater, built northwards from the mainland to Chapel Rock to protect the harbour from the battering of the Atlantic waves. Many years before the breakwater was built, until the harbour became silted, spring tides would reach more than a mile inland from the haven mouth and ships came right up to the old quay by the Strand. When the accumulation of sand made this impossible ships were beached on the sand under Summerleaze. The building of the breakwater and the alteration of the course of the river from its former more northerly channel below Summerleaze changed the topography of the harbour and protected the important reserves of sand as well as making the mouth of the river navigable at high tide. The first breakwater, in spite of its later-proved inadequate strength, was said to have been 'much finer looking' than its successor; 900ft long, about 20ft high and

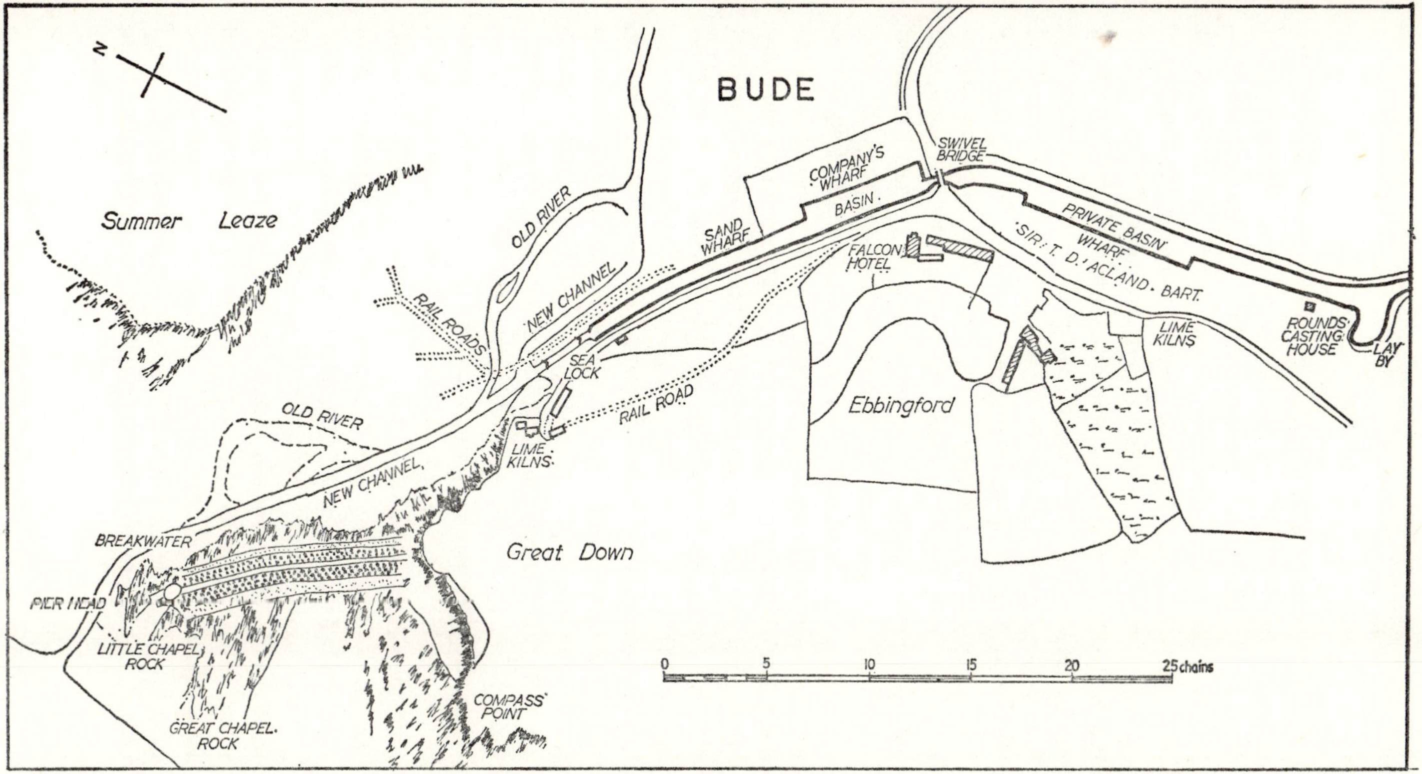

FIGURE 3. The canal in 1826

tapering to an upper width of 10ft, in section it was 4ft taller and more slender than the later one, forming a hollow curve on the seaward side and terminating abruptly in a granite pier-head surmounted by a circular tower at the north end. It was faced throughout with well-finished blocks of granite and limestone, cut and dressed in quarries at Plymouth and Penryn and in Wales, and was composed internally largely of clay and earth.

The sea lock, construction of which necessitated cutting the Shaldern sandhills which previously reached from Ebbingford Manor to Nanny Moore's bridge, was recorded by Thomas Telford in his notebook as being constructed 38yd in length and 11yd in breadth. Forming the entrance to the canal, and with a depth of about 15ft at average spring tides, it enabled ships of up to 300 tons to enter the basin, which, for the first ¼ mile, was 10ft deep. Besides the docking here of sea-going vessels and the transfer of cargoes where required between them and canal craft, it was at this point that barges and tub-boats took on their loads of sand. Instead of Green's original plan for floating sand barges into the lock on the rising tides (see p 43) Lord Stanhope's suggested modification (p 44), involving the use of a narrow-gauge railway and special trucks, was the method adopted for the transfer of sand from the beach to the quay-side. Rights of the higher of the basin's two wharves were reserved by the lord of the manor, Sir T. D. Acland, though full use by the canal company was allowed.

Beyond the main basin the canal was shallower, with a depth of 4ft 6in for the first 1¾ miles to Helebridge, and navigable only by the barges and tub-boats. Two locks on this stretch, one at Roddsbridge and another at Whalesborough, both 63ft long by 14ft 7in wide, each raised the level a further 5ft 6in.

The barges, which carried up to 50 tons and measured 50ft by 13ft by 3ft 6in, could travel no farther than Helebridge wharf. Here the goods, mainly sand from the beach or coal from South Wales, were transferred into tub-boats for the remainder of the journey. The wooden tub-boats, some of which, as already mentioned, plied direct to the main basin,

Page 51 (*left*) *Traly* entering the sea lock at Bude; (*below*) Sailing craft and tub-boats moored in the canal's lower basin during the latter part of the nineteenth century

Page 52 (*above*) The lower basin at Bude filled with craft towards the end of the last century; (*below*) The upper basin at Bude showing Falcon bridge before it was fixed. In the right foreground can be seen the railway line of the siding which connected the lower wharf with the station

measured 20ft by 5ft 6in and were each capable of carrying up to 5 tons; they were drawn by horses in sets of four or six boats strung together, the leading boat having a pointed bow. A man stood in the bow of the second boat in order to steer, which he did by means of a handspike applied between the first and second boats to maintain or alter the alignment as necessary. Projecting from the sides of each tub-boat were four cast-iron wheels of 14in diameter which came into use, running in rails, as the boats travelled over the inclined planes. An iron bar was fitted at the stern of each boat which, in the event of a chain breaking or a mechanical breakdown occurring on an inclined plane, would be forced into the ground in an attempt to prevent the boat running backwards.

The first of the canal's six inclined planes to be reached was at Marhamchurch, 2 miles distant from the sea lock. This plane, the longest but one, raised the level a further 120ft in a distance of 836ft. Motive power was provided by a 50ft diameter, 3ft wide overshot waterwheel situated at the top which operated an endless chain over a winding drum to draw the boats up the incline; 32 tons of water were required to raise a 4 ton boat and the time taken for passage up the plane, including starting and stopping, was five minutes. All the planes had double sets of rails, dipping into the canal at each end, so that boats could be raised and lowered simultaneously.

From the top of the plane, where boats were refloated into the next level of the canal—of 3ft depth, 10ft wide at the bottom and 19ft at the top—the route passed immediately north of Marhamchurch village and followed a somewhat winding course north-eastwards a further 2 miles in the direction of Thurlibeer. Here, at Hobbacott Down, was the biggest and most ingeniously impressive of the inclined planes. Rising 225ft, 935ft in length, the Hobbacott plane, like the others, was worked by water power, but not by means of a waterwheel. The system here involved two wells sunk at the top of the plane, corresponding in depth to its vertical rise, in each of which the descent of an enormous bucket, 10ft in diameter,

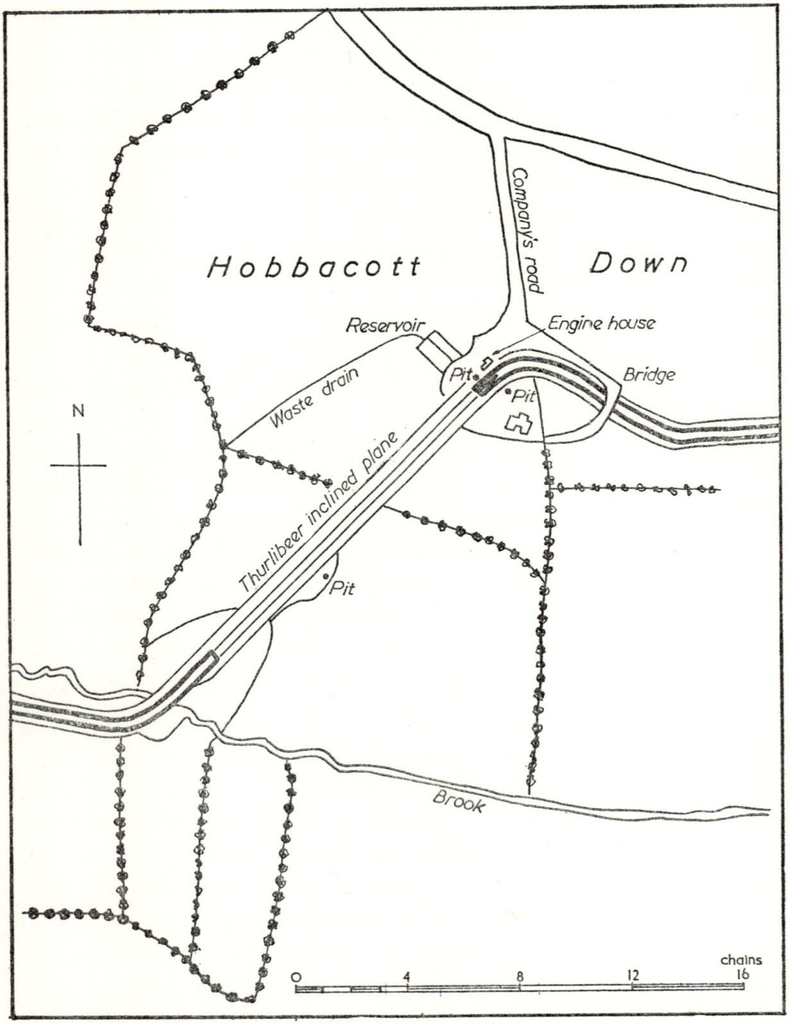

FIGURE 4. Hobbacott inclined plane in 1826

5ft 6in deep, suspended on a chain wound over a drum and containing 15 tons of water, provided the power for raising the boats. As the bucket completed its descent and hit the floor of the well a valve in its base opened and allowed the water to escape along a drainage channel to the canal's lower level, the second, empty, bucket in the other well meanwhile having been raised to the surface ready for use again. As a standby for emergencies a 16hp steam engine was also kept at the top of the plane and was, in fact, not infrequently in use, since break-downs, generally due to a chain breaking, were quite common; when this happened to the bucket chain it was said that the reverberations in the pit caused by the bucket falling to the bottom were 'terrific and if once heard never forgotten'. Loaded boats ascended the plane by bucket power in the astonishingly short time of four minutes, double that time being taken when the steam engine was employed. There was, however, a water-wheel in use at Hobbacott—a small one of 9ft diameter worked for providing power in the canal blacksmith's shop.[1]

From the top of the Hobbacott plane the canal's main trunk ran almost due eastwards for a further 1¼ miles to Red Post. Here it passed beneath the Kilkhampton–Launceston road and immediately afterwards divided into its two branches. The Holsworthy branch, of about 10 miles, continuing generally in the same main direction, first took a wide southerly loop around Shernick and then passed beneath the Holsworthy–Bude road and around Anderton to Burmsdon. At this point the canal crossed the River Tamar by a fine stone single-arched aqueduct, entered the county of Devon and approached the next inclined plane. This plane, known variously by the names of Vealand, Veala or Venn, was worked by a waterwheel and raised the canal a further 58ft in a distance of 500ft, taking it to the summit level. On Brendon Moor, a short distance from the top of the plane, there was a junction on the north side with the feeder branch from the reservoir, while the canal itself continued its winding way through the parish of Pancrasweek and south of Chilsworthy to its destination. At points along this stretch, and in particular where the canal passed beneath the Bradworthy road, north of Burnard's House, deep cuttings were necessitated. There was a wharf near Stanbury Cross (north of the town of Holsworthy), where the canal passed beneath the Holsworthy–Bideford road, and another just over a mile farther along, at the Blagdonmoor terminus.

The feeder branch meandered from the junction at Brendon for about 4½ miles to the reservoir at Alfardisworthy, serving a wharf en route near Virworthy for the convenience of farmers in the parishes of Bradworthy and Sutcombe. The 70 acre reservoir, of 195 million gallons capacity, sited in a remote upper reach of the River Tamar on the Cornwall–Devon border, was formed by the construction of a dam on the southern side and fed by the river's flood water and by other minor streams.

At Red Post the 16 mile Launceston branch made a quite sharp turn away to the south and kept to the west of the River Tamar for the whole way. The first wharf reached on this section was at Littlebridge, Bridgerule, just west of the village; past it the canal crossed the minor road by a wooden aqueduct, then turned to run beneath the Bridgerule–Borough Cross road. The first of the branch's three waterwheel-worked descending inclined planes was situated 4 miles along the line from Red Post at Merrifield and had a fall of 60ft and a length of 360ft. From here the route, following the contours of the west side of the Tamar valley, continued generally south-eastwards to North Tamerton, where a wharf was sited at the point where the canal crossed the road from Tamerton bridge to the village by a rather low aqueduct. A little over a mile farther on and 7 miles beyond Merrifield was the Tamerton inclined plane, also of 360ft in length with a descent of 59ft. After a further 4 miles, during which it served a wharf at Boyton bridge and crossed by an aqueduct the minor road at Tamartown, the canal reached the last of the inclines, the Bridgetown or Werrington plane, which had a fall of 51ft and in its length of 259ft passed over the slightly lowered roadway by means of a stone arch. In another mile, crossing the road again, the canal reached its southern terminus, known as Druxton wharf, but actually at Crossgate, 3 miles north from the town of Launceston.

FAULTS AND FRUSTRATIONS

The whole system was well planned and exceedingly ingenious, and when all was in order worked extremely well. Unfortunately, however, the inventiveness was not fully matched by much of the actual construction and the materials used, with the result that breakdowns were frequent, particularly on the inclined planes. These involved not only delays and resultant loss of trade but considerable expense in the cost of repairs which far exceeded previous estimations. That the trouble occurred from the very earliest working days is evident from monthly reports on the canal works made to the Bude Harbour & Canal Company by the then clerk, John Honey, and by George Call who, in spite of the earlier differences of 1822–3, was at this time resident chairman of the committee of management. The report of 2 May 1827 when the canal had been worked in its entirety for six months must have made depressing reading; in it the chairman 'with the deepest regret' called the committee's attention to the contingent account amounting to £2,747 18s 5d (£2,747.92), a sum which exceeded the amount of tolls for the year.

The types of fault in the structure were varied. In November 1826, for example, it was reported that the greater part of both the up and down tram rails on the Marhamchurch plane were in a very bad state, causing strains to machinery and injuries to boats, the metals being so broken in short lengths that it was impossible to keep them in one direction; and the circular wood part of the horizontal chain wheel at the foot of the plane was found to be rotten. At Hobbacott and at Vealand also there were rails in a very bad state, similar to those at Marhamchurch. Leaks and slipping banks occurred frequently, and, particularly on sections of the Holsworthy branch where embankments crossed valleys, there was trouble with sinkage; the banks were made up from time to time but the bed of the canal tended to drop below the required level. Sometimes the boatmen were to blame for not steering with sufficient care—on one occasion Mr Call felt obliged to caution them—and in May 1827 he recommended that at Red Post, where the southern line branched off by a short turn, a stone wall and coping should be built to counteract the severe friction of the boats. Often, if the boats were light and there was a breeze, control was difficult and banks were damaged by the projecting wheels of the boats as they scraped against the sides. (By 1838 the wheels had been enclosed on each side within a 'closed trough' so that they ceased to project, except slightly below the lower surface.)[2]

The breaking of chains of the inclined planes was one of the biggest nuisances. One instance was on 14 December 1826 when both chains of one of the buckets at Hobbacott broke and the bucket fell from halfway down the pit to the bottom. Damage was greater, however, in a similar occurrence of 5 May 1827 and necessitated repairs estimated to cost at least £20–£25. Breakages of the actual drawing chains too were not infrequent; in that same May there was a break in the chain at Hobbacott causing a sand-filled boat being lifted to descend rapidly, crashing into and damaging two waiting boats at the bottom, a break in the main chain at Vealand—fortunately with no damage—and another parting of the main chain at Hobbacott. And at Tamerton plane there was a break in the shaft of the chain wheel while a boat was being let down.

John Honey had been appointed clerk to the company at Bude in 1824 and remained in the company's employment at least until March 1832. During these years he moved into a house at Hobbacott Down, at the top of the plane, succeeding a Mr Whitewood (who later went to employment in the construction of the Grand Western Canal) and apparently acted as agent and shouldered many of the engineering responsibilities in addition to his clerical work. He was, in fact, highly regarded by many people in the district, including George Call, who backed him to the committee on several occasions. In a letter accompanying one of Honey's reports, in 1826, Mr Call wrote: 'I am so much satisfied with the clear and comprehensive manner in which these documents are drawn up that it would be a waste of your time and an act of Injustice to Mr Honey to attempt any alteration or addition thereto. I agree with him on all points particularly on the necessity of repairing the Railways on the Planes.' The demands faced by the clerk and the hardness of his work were recognised by the resident chairman when he further reported to the committee in January 1827. 'The last month has been a period of misfortune(s) and they have followed each other so quickly that Mr Honey and his Assistant have had most severe duty to perform. I leave you gentlemen to judge of the exertion of the former when I state that he rode three hundred miles in five days and I am astonished at his having been able to stand up against such fatigue. I trust Gentlemen that you will allow me to suggest the propriety of noticing his good Conduct in your Resolutions.' And in a letter to the same body in May he stated: 'I beg leave to assure the Committee that in every Branch of the Service connected with the Canal Office I have derived the greatest assistance from Mr Honey—his real Activity and Intelligence on all occasions entitle him to my best thanks.'

Nevertheless there were those who failed to appreciate the difficulties of the situation and in particular it appears that the committee, which generally met at Exeter, did not always give its men on the spot the support they deserved. From a series of lengthy letters written by John Honey between 29 October 1830 and 27 March 1832 to J. Blackmore and J. W. Crabb, both of Exeter and chairmen in turn of the committee of management, and to others including T. Shearm, the company's land surveyor, it is possible to form a picture of events. Life for Honey meant coping with the frequent breakdowns, accidents and damage which occurred and trying to find sufficient money for paying taxes, to pay tradespeople for materials and spare parts supplied, labourers for their wages and various individuals for occasional incidental damages. This was not easy since trade on the canal fluctuated and the traders who operated were often dilatory over the payment of tolls, and obtaining cash from the company was a prolonged procedure. There were difficulties in getting answers and decisions, and delays over repairs while these were awaited and until overdue parts were received.

Extracts from some of the letters will give an idea of the problems with which John Honey had to contend. Concerning the canal itself:

> *23 December 1830* 'I beg further to inform you that the connecting shafts between the waterwheel and the Chain Wheel shaft at Tamerton Town plane broke on Tuesday last in taking up some returning boats over the Plane, I went on yesterday with Geake and investigated the particulars and I am making a Model for a Cast Iron Clip to mend the shaft, and hope to get it cast on tomorrow & fixed up on Monday next, when it will be in proper order for the Trade again if there should be any.'

> *9 January 1831* 'I informed the Committee when at Exeter that the Wheels at Tamerton, Merrifield & Bridgetown Inclined planes wanted new hanging. The men (mechanics) of these planes being all discharged I cannot do it, therefore I hope you will attach no blame to me in case of accidents when the work should again recommence; I also informed the Committee that a quantity of fences throughout the line was necessary to be done, having only 5 labourers in 35½ miles of Canal this cannot be attended to all throughout, great trespass of cattle will be the result.'

> *11 February 1831* 'I am extremely sorry to inform you that the Boiler of the steam engine is rusted through in many places, which prevents us from working at present in any way at this place [*Hobbacott Down*].'

> *6 April 1831* 'I have examined and repaired Marhamchurch Machinery so far as I can (viz) in putting all the Gear to run correctly, and have put new bearing Brasses under the Water Wheel shaft, and rose it up to its correct level and height. This shaft had worn through the bearing Brasses and working on the cast iron carriage, which was highly improper. . . . The wood framing is in many places extremely rotten, something must be done to it soon, otherwise the iron machinery which is

at present supported by it, must be damaged considerably, and if worked to any extent the time will not be far distant when these damages must occur.

'I have thoroughly examined the state of the machinery of all the small Inclined Planes as to the Geer, the manner in which it now runs and also the wood framing that supports it; there is much wanted and actually necessary to be done, both, as to fixing the Geer more correctly & supplying new framing in a great many places, where the old is rotten and defective, if not soon attended to provided that the Canal again revives to a full state of work, the damage which must occur (and no one knows how soon) is at present beyond calculation.'

'I am repairing the Steam Boiler and shall lose no time in putting it in proper repair.'

'The Valleys throughout the line generally want a thorough repair most particularly the large valleys viz Pancrasweek, Chilsworthy & Manworthy. . . . The sides of the Canal or otherwise the top parts of the banks on each side these valleys have been continually risen upon, as they have sunk in the center, to keep them above high water mark, and the bottom has never been brought up to it, and worse than all I have before represented is, that your late engineer has actually built cob (or mud walls) on each side of the valleys next the water, for the purpose of strengthening them. This I cannot help calling *Engineering to perfection*, the poor labour men who were obliged to do it under his instructions told him that they knew too well the nature of cob walls (from their own cottages) for it to answer any purpose wherein it was kept in the least degree wet, much more so, where it was always under water his answer was to them (as in all other cases) let it be right or wrong do as I tell you, then you'll have no fault, and when they would say they thought it would not answer, he would say in answer *d——m thee*, thee has no right to think, *I must think for I am paid for thinking*, and thee for working, therefore do as I tell thee or be off.'

28 June 1831 'It is necessary the Steam Engine be immediately put in repair, in case an accident should happen at Hobbacott Down to the Bucket Machinery.'

13 August 1831 'About half past twelve o'clock this day a man named John Wilkey a boatman in Mr King's employ in the act of taking down some empty boats from the Top of Hobbacott

Plane, hooked on three nos 30, 45 & 46 and put the shackle on the Main Chain in the regular way; immediately as the two foremost boats came out over the Top of the Plane, this man Wilkey who was in the foremost boat was seen by Wm Dyer a mason working for the Company removing or handling the Shackle; and by this means the Boats escaped from their fastening & run from Top to bottom of the Plane; on examining the foremost boat, I found the V which the Main Chain falls into also broken as they came to the bottom the foremost boat ran immediately into the center of the Work and Broke the shaft of the Chain Wheel & several of the Wood arms & has considerably damaged the framing. This accident and damage has been occasioned through the mismanagement of the Boatmen. The Companys men were working the Machinery in the regular way. I was in the office at the time on the Top of the Plane in time to see the Boats arrive at the Bottom. The damage done will occasion the Trade to stop for 10 days or a fortnight. I can repair the Chain Wheel shaft without having a new one. The framing which is so much damaged was much rotten in some places, new Timber must now be had to repair it. . . .'

8 September 1831 'It will be my greatest pride to put and see the whole of the Machinery in good order and the Trade in a flourishing state, and nothing ever shall be wanting on my part to effect it, but from the present appearance of repairs which in justice ought to be done to the Machinery at Marhamchurch, Hobbicott Down, Vealand & Merrifield Planes and the expense always attending that at Hobbacott, that I have some reason to fear from the extensive and complicated Works at the latter place, it will always be a Mill Stone on the Companys neck. I well know it is now no time for reflection but had this plane been fortunately made into two, in a different direction and worked by two water wheels, almost all the former enormous expense, and that which in future must be, would have been saved.'

5 October 1831 'On Monday afternoon the connecting shaft from the Water Wheel at Bridgetown Plane broke in two, . . . no damage was done to the boats but the small bevel Gear Wheel which form a part of the Gear for hauling Boats from the Canal to descend was also broken.'

4 February 1832 'Hobbacott Down Machinery . . . was very much damaged on the 1st Inst by the Main Chain breaking.'

With regard to finances Honey wrote:

3 November 1830 'The last small Traders are selling Sand without any profit to themselves and is the means of very considerably injuring the other Traders. . . . Two months wages being due to the men to 31st Oct & many of them discharged it is necessary they should be paid. I have called on Mr King and Mr Bray this morning and cannot get a shilling. Mr King says he may be enabled to get some on tomorrow, Mr Bray says it is impossible for him to pay monthly, if he is pressed to do so, it will be his ruin and he must give up the Trade immediately.'

20 November 1830 'believe me Sir, I assure you that I am almost hourly annoyed here, by being called on for Labourers Wages Taxes ETC and not a farthing have I to pay either. It appears to me that the Farmers who are the only support of this Canal, have by some means totally set their face against it.'

23 November 1830 'It being now Market Day at Stratton where the Mechanics & Labourers get their supply for the ensuing week I have had almost the whole of them . . . requesting to be paid their wages, as otherwise they inform me they cannot procure necessaries for their support. . . . Messrs H & J James have applied to me for payment of the amount due to them for Sundry Goods . . . I have been called on this morning by no less than 10 persons for Taxes. I've not a farthing to pay them, and worse than all this there is no Trade whatsoever on the Canal and only 177 Boats have been taken over the Planes from the 1st of this month to the present time. This Trade will not pay for Coals and Labour in passing the different planes. . . . The farmers are going to the Beach for sand & to the Lime Kilns for Lime. . . .'

16 December 1830 'If there is no trade there can be no income, and I cannot compel the Traders to work. I've done all I can both with the Traders and farmers and the former informs me they cannot sell their sand if they take it up, if they are to pay the Tolls now demanded, the latter informs me they cannot purchase sand at the price the Traders sell for, . . . I am confident in my opinion that unless some alteration takes place immediately on the Tolls that the Company will have no money even to pay the Taxes on the Canal . . . you must be aware that the advance of Tolls has been the only means of stopping the Trade. . . .'

6 April 1831 'Where am I to get money to pay the labourers etc. £136. 16. 1¾d being now due.'

20 July 1831 'What am I to do for money to pay the Labourers Taxes etc. Capt King has paid me his last months tolls which I have distributed among the Labourers etc. I have called on Mr Bray and his answer is that he shall not be able to pay any tolls before Michaelmas next and Mr King informs me that he will pay this months Tolls when it becomes due; after that, he shall not be enabled to pay monthly in full before Michaelmas. I beg leave to inform you that in case the Labourers cannot be paid their wages monthly, that they cannot get necessaries for their support without paying extravagant prices. . . .'

It was no wonder that Honey felt tired. Supervising the whole length of canal, dealing with emergencies and effecting repairs—often with his own hands—endeavouring to manage the finances and reporting to the chairman must have taken every minute of his time. Yet even when he wrote to the chairman 'I have walked myself until I am almost tired in that degree that I can scarcely walk any more' the committee still would not provide him with a horse, and when he himself borrowed one it could not finance a saddle and bridle. Early in 1832 Honey was given notice to quit, ostensibly as a measure of economy, and obviously felt resentment at his summary dismissal. He was succeeded in the superintendence by one of James Green's assistants, a young man called Joseph Cox, but may have continued in the office of clerk until 1834, when James Collom was appointed.

FINANCIAL CRISIS

During the early 1830s the shaky structural condition of much of the canal and the hard-pressed financial state of the canal company was a matter of concern to the Exchequer Loan Commissioners. During the late 1820s the company had been faced with considerable expenses; sea damage had necessitated repairs to the breakwater costing well over £1,000 and there had been endless breakages of machinery and other faults to correct, besides which there were still debts for land purchased which amounted to around £2,000. From 1827 the company had been unable to repay the required instalments of principal on the exchequer loan and in 1829, after various interim indulgences by the loan board, a three-year period of postponement was asked for and granted. Though a succeeding interest payment was made in September 1830 there had been no further repayments of principal, and on 19 June 1834 a memorial from the Bude Harbour & Canal Company was read at a meeting of the Exchequer Loan Commissioners. The communication expressed the company's hopes that by alterations of toll and by strict economy of expenditure all debts would be liquidated within the originally allotted twenty years, but revealed that further expensive breakdowns of canal machinery had occurred —particularly concerning the workings at Hobbacott where it had been found necessary in the previous two years to erect new machinery—and that trade had consequently been greatly impeded and considerable expense incurred which had absorbed all the company's receipts. Remedial measures, however, were being taken: modification or a change of machinery at the Hobbacott plane were under consideration, an increase of toll which had been tried without success had been replaced by a test of the effects of a decrease which had proved contrastingly encouraging, and the meetings of the company were being transferred from Exeter to the immediate vicinity of the canal which was expected to ensure prompt and immediate attention to the interests of the company. Whilst declaring itself open to any suggestions as to arrangements or economy, the memorialists appealed to the board for suspension of immediate payment, undertaking to pay future instalments as due and to discharge the whole of the balance within the agreed twenty-year period.

The Exchequer Loan Commissioners' secretary, having been invited to examine the works, subsequently journeyed to Bude and made a close examination of the affairs of the company in an attempt to discover the cause of the financial embarrassment and to seek a remedy. His opinion affirmed that the principal impediment to the canal's success was the Hobbacott plane. Where there might have been a simple and efficient contrivance for raising the boats he saw instead Hobbacott's complex method of water buckets and multiplying wheels and learnt at first hand of the repeated serious accidents and the expense caused, of the lack of confidence in the traders and consequent loss of revenue, and saw the state of dilapidation into which the works had fallen. On the day of the secretary's visit no less than eighty loaded boats were stopped at the great inclined plane and his conversation with the traders produced the uniform opinion that the traffic would greatly increase if only a proper facility of transport were given; he was convinced that if the prevailing system continued during the lapse of time requested, the canal affairs would run into an even worse state. On consideration the board felt that there was no reasonable prospect of loan repayment under the existing expensive and inefficient mode of operation. The necessity for a more efficient and economic method of passing boats over the principal inclined plane and for other improvements in the company's proceedings were pointed out to the committee of management and the committee was advised, in the event of the Hobbacott machinery again giving way, to try the effect of material alteration. To such an improvement the board was disposed to offer facilities in the repayment, and to defer for another year the claim for arrears, but in the meantime required the committee of management to render monthly accounts of its proceedings.

In January 1835 the engineer James Meadows Rendel was engaged by the canal company's committee of management, with the sanction of the Exchequer Loan Commissioners, to report on the state of the works on the line of the canal. Rendel's report indicated these to be generally out of repair and likely to get rapidly worse unless a considerable sum was promptly laid out in restoring them, and that, in his opinion, the principle of construction of the bucket machine at Hobbacott was not such as could be depended on for carrying on the trade. In a letter the following month, while emphasising that in spite of the difficulties trade on the canal had improved and was anticipated to increase further, George Call, on behalf of the company, formally requested the board's assistance in meeting the difficulties and asked for a further advance of £10,000, on the guaranteed security of certain individual shareholders who would make themselves responsible for the entire management of the work until the sum was repaid.

No immediate action was taken by the board and there followed a voluminous exchange of documents and letters between individuals, suggestive—though precise contents are not known—of conflicting elements and possibly even of intrigues between members of the canal company. James Green gave professional observations on Rendel's report and entered into a correspondence with the Rev N. Cole, a member of the committee, who in turn communicated with the board's secretary. The committee of management, under the chairmanship of Sir William Pratt Call, moved a proposition at the general meeting of proprietors to obtain an Act of Parliament for extending the powers of the company to raise funds, but the proposition was negatived and the committee promptly resigned. A new committee was then chosen, and the Rev Cole elected chairman.

The new chairman made instant application to the board, on 9 May 1835, for 'still further extension of that forbearance which the Hon. Board has so frequently exercised' and for permission to apply the current income to the payment of the committee's land and other outstanding debts, and afterwards to apply the surplus income to the liquidation of the public debt. To this the reply was 'that the Board do not feel themselves disposed to sanction any appropriation of the accruing Income of the Undertaking to the Extinction of the Company's Land Debts and other outstanding debts'. It appeared that the new party in power had decided not to adopt a recommendation by the former committee that a powerful steam engine should be substituted at Hobbacott for the bucket machinery and small steam engine, but had contracted with James Green to repair the bucket machinery at a cost of £300, to work and keep it in repair for one year for £260 and to spend £100 on reconditioning the small steam engine. This arrangement, which took no account of other immediately necessary repairs and contemplated the working for one year only, offered little prospect of the liquidation of the public debt, and Green's intention to incur expense in retaining the small steam engine indicated a lack of confidence on his part in the stability of his bucket machinery. Taking into account Rendel's report the board did not consider the proposed measures acceptable, as they afforded no expectation of a permanent security of working, and the company was informed that, unless some arrangement could be made for the payment of arrears due, the commissioners would consider it necessary to take possession of the undertaking in order to realise the monies.

This ultimatum provoked considerable anxiety when read at the next meeting of the committee of management, and, with the Rev Cole in the chair, it was unanimously resolved

> That the Committee feel deeply concerned that the Hon Exchequer Loan Board should at a moment when the Trade on the Canal has increased and is increasing to an extent of nearly double that of any former period, and when there is every prospect of the Canal Expenditure being greatly reduced proceed to such decisive steps as stated in their letter. It can be fully proved that the purpose for which the Canal was Cut and the Improvement of the Harbour of Bude, have been fully verified in as much as the Landed Proprietors and others in the neighbourhood have increased the value of their Lands from 30 to 40 per cent and in some parts still more, whilst the Canal has afforded constant employment to the labouring population: it is well known that the Landed Proprietors were unable to cut the Canal without the assistance of the Monied Proprietors in the neighbourhood, who hold shares to the amount of nearly £45,000 without having yet received any benefit whatever; but have on the contrary sustained a total loss of Interest and will if the Exchequer Loan Board carry their intention of selling the Canal and Works into effect have a great reason to fear a total loss of their capital, whilst the Landed Proprietors (who will

Page 69 (*above*) The present breakwater, constructed 1838–43. The previous breakwater, a slightly higher and more slender structure, was destroyed during a great storm in 1838; (*below*) The massive entrance to the sea lock. The mooring post on the left marks the channel. The bridge behind it carried the sand railway up beside the sea lock to the lower wharf

Page 70 (*above*) Looking out through the sea lock gates towards the end of the breakwater showing the Chapel Rocks and the line of the channel; (*below*) One of the storehouses on Sir T. D. Acland's wharf above Falcon bridge

most likely become purchasers) will reap a double benefit by making the purchase at a low rate and benefitting their Lands to a still greater degree. The Committee of Management feel confident that should the Exchequer Loan Board be disposed to extend their indulgence, the measures they have in contemplation and which they have submitted as contained in the Resolutions of the General Meeting on the 3rd and Committee resolutions of the 5th will be the means of extricating the concern from all its difficulties, particularly as the low state of Toll will at no distant period admit of being raised at least $\frac{1}{8}$. The Committee have also ordered the repairing of the Machinery at Hobbacott Down to be proceeded with by a Resolution of the 3rd instant and which they now understand to be in a forward state.

The resolution was sent to the Exchequer Loan Board which was evidently impressed by the committee's confidence that the measures would extricate the concern from all its difficulties, and it relented to the extent of granting another year's indulgence, provided no further expense should be incurred beyond that already referred to, and that should the bucket machinery and steam engine fail to act after the repairs at any time during the year, the board would feel it its duty to take the proceedings previously intimated.

From the absence of adverse reports it appears likely that the immediately subsequent running of the canal went reasonably smoothly, though in 1836 the board was again asked for an extended indulgence over the payment of interest. The canal company was evidently doing its best to consolidate its position and no doubt a revival and strengthening of trade provided the board with an indication of steadily increasing security.

TRADERS AND EMPLOYEES

Four companies were trading on the canal in 1838, those of H. King, S. Bray, H. Rundle and B. Adams. Each had to pay a certain sum of money to the company as security. Toll amounts were adjusted from time to time, but these were generally higher than the traders liked if they were to offer the sand at a price sufficiently attractive to the farmers. To keep them operating the company had to lend capital to the traders, not being permitted under the Act itself to trade on the canal. Although, with the exception of King, the traders were often overdue on their payment of tolls the traders were commended by Mr Call for the cheerful way in which they submitted to unavoidable stoppages of trade. Early in 1832 Rundle sold his boats to King and went to America, and in 1841 Captain King, RN, himself left the country on being appointed Chief Commissioner to the New Zealand Company, an occasion which prompted the committee of management of the Bude Harbour & Canal Company to send him an address, noting the 'honourable zeal' with which he had 'so ably conducted the principal mercantile establishment of the Bude Canal'. Prior to his leaving Capt King accepted an offer by the company of £150 for the sale of a cottage at Druxton wharf and coal sheds and stables at Stanbury Cross and Blagdonmoor, which the company subsequently let to Ham & Co. At this time, the early 1840s, the company was improving facilities at the wharves with the addition of coal yards and buildings; individual traders had their own allotments of space at the wharves, identified by their initials on iron stakes. In 1841 the firms of John Somers James & Co and the Bude Canal Trading Company were operating on the canal.

The men who worked on the canal, mechanics and labourers, were also praised by George Call when in the early period they had at times worked day and night on repairs. Nineteen men were employed in 1830, consisting of a lock-keeper, wharfinger, mechanics and others on the planes, masons and men who maintained the banks and fences. Later, in the 1840s, it is said that the canal provided direct employment for over 100 persons. The company occasionally bestowed merits; in 1839 it resolved 'That Thomas Smale employed at Tamerton be paid 2s instead of 1s. 8d. per day from this date as a reward for his attentive conduct to his Section of the Line being in a better state than any others and being strongly recommended by the Engineer'.

And in 1841 'That the sum of ten shillings be paid to the labourer whose portion of canal shall at the expiration of one year appear to Mr Vowler to be in the best order'.

WEATHER HAZARDS AND A FURTHER CRISIS

Naturally enough, extremes of weather occasionally affected the working of the canal. When there was drought and the reservoir dropped below working level, and with the need to allow sufficient water to pass for mills downstream, the canal had to be temporarily closed; it was also generally closed when harvesting was in progress and advantage taken at such times for doing any necessary repairs. In severe frost the canal could freeze and trade would be suspended; to overcome this experiments were made in January 1841 with an ice-boat, constructed by George Casebourne (engineer in succession to James Green), drawn by two horses and travelling at 4mph, which broke the ice several inches thick.[3] Floods were also a hazard. In the early years it was necessary on one occasion of excessive flooding to cut through the banks in places to let off surplus water; after this trunks and paddles which could be lifted by rack and pinion were provided for use at such times and in case of accident. In May 1831 sand was lost at Helebridge and Blagdonmoor wharves by the sudden swell of flood waters.

The breakwater suffered most from the natural elements. Breaches due to heavy seas and rough high tides occurred at intervals from the earliest days and necessitated frequent repairs, generally hurried between tides, by Honey and his men. One fault seemed to be the lightness and irregular working of some local Compass Point cliff stone used in the original structure (replaced with Bristol stone in repair) and the filling of loose rubble and earth which could be washed out by a strong tide. A climax came on 24 and 25 February 1838 when there was a terrific storm, more violent than for years. Damage occurred to sea structures all along the south coast of England, including the Plymouth breakwater, and the breakwater at Bude, exposed to the full force of the Atlantic, was a ready victim. The storm began in the evening, with the wind in the south, through the night it veered through SW to WSW and it was at 5.30am, half an hour before high tide, that the breakwater collapsed. The inside slope gave way first, due apparently to the heavy waves rising up the sea slope and over the top and falling with great force on stonework whose mortar had been weakened by frost a month previously, and on the internal rubble. Very soon three-quarters of the breakwater was destroyed to a depth of 12ft and the material of which it had been made deposited over the shore.

With the company's outstanding debts—over £24,000, including the £21,037 loan and interest due to the Exchequer Bill Loan Commissioners—and with the non-payment to date of any dividend, financing the breakwater's reconstruction presented a problem. But trade on the canal had improved. Though the early estimates of sand consumption were never realised a considerable traffic had built up and in the year ending 31 March 1838 the combined canal, harbour and basin dues amounted to over £4,300 (sand formed nine-tenths of almost 60,000 tons of goods carried), expenses being about £2,600.[4]

In view of the surplus the company decided to ask the Exchequer Bill Loan Commissioners for permission to use this fund for the breakwater instead of for interest payment; the Commissioners instructed James Walker to make an inspection and report on the system, and on his recommendation that the breakwater was essential and the canal beneficial to the country through which it passed, approval was given for the surplus income to be applied to the work for three years without paying interest.[5] Work proceeded over the next four years, using stone salvaged from the first structure and local stone supplied by Sir Thomas Acland which was brought from under and beyond Compass Point. A tramway and timber road were used for transporting the stone, with the enormous pieces of rock hauled not by horses but by men—generally twelve men being required to drag each piece—and a whim was employed for

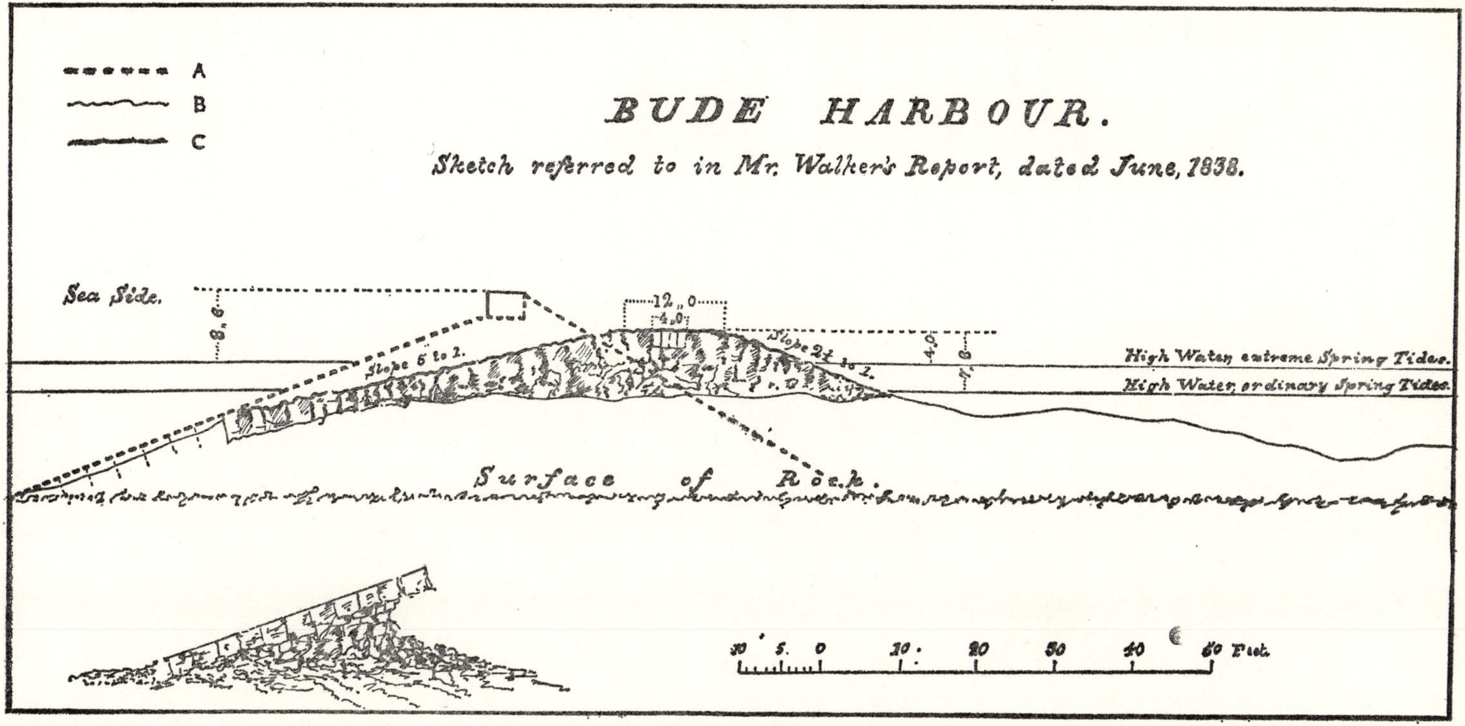

FIGURE 5. Sectional sketch of Bude Harbour 1838
A, before the storm; B, current state; C, proposed repair

drawing the waggons up the slope. The completed breakwater was lower and more gently sloping and of rougher appearance than its predecessor; though in later years yet further reconstruction work was necessary it has in general proved sufficiently stable to withstand the force of the seas. The company's resident engineer by the time of the rebuilding was George Casebourne, an able man who had been in the company's employ since 1832 and remained in the appointment until 1876.

In 1839 approval was given by the Exchequer Bill Loan Commissioners to another application from the canal company, for permission to extend the wharf at Helebridge, at an estimated cost not exceeding £90.[6] In this year the debt to the Exchequer Loan Board stood at its highest figure of £22,427, including interest. From then on, however, the situation steadily improved; in 1841 income from tolls reached its highest level, of £4,557, in 1848 the debt was £15,731, by 1864 it had been reduced to £3,000 and in 1870 it was finally paid off. Although for so long the canal remained a disappointment to its subscribers it had proved of advantage to agriculture; farmers were enabled to buy sand at a much cheaper price than previously and thus to be more liberal in its use, resulting in a noticeable improvement in the tilth and an increase in the rental value of land along the whole of the canal's length.

CHAPTER 4

Later Years: Decline and Demise

Following the boom years of the early 1840s the canal worked its way through the 1850s and into the 1860s with a general decline of takings—partly due to improvements in road transport—but without any major obligations of structural expenditure, and with a steadily strengthening financial position as the debt to the Exchequer Bill Loan Commissioners was being paid off; 1859 was the year which showed the highest profit, of £1,493. An innovation introduced in about 1860 was a steamboat, purchased by a shareholder, Mr S. Smith of London, which was employed for drawing the barges from Bude to Helebridge and as a tug-boat for assisting vessels in and out of the harbour; its use soon ended, however, as it was found to be damaging to the canal banks and uneconomic in operation.

RAILWAY INGRESSION AND ITS EFFECTS

In the 1860s the Bude Canal felt its first actual effects of railway development. There had, as far back as 1836, been the possibility of severe competition when an Act was obtained for making and maintaining a harbour and breakwater at Tremoutha Haven, a few miles down the coast at St Gennys, with a new town there to be called Victoria and a railway connecting it to Launceston; but, fortunately for the canal, the plan never materialised. But in 1864 railway advance into the district was a reality when the Launceston & South Devon Railway was opened, connecting Launceston with Tavistock and Plymouth. At first there were hopes of advantages when in 1865 an Act promoted by three Launceston men was obtained to incorporate the Bude Canal & Launceston Junction Railway Company, with a capital of £20,000. The Act authorised construction of a 2½ mile railway from the Druxton basin to two junctions with the L & SD line—one for Launceston and one for Tavistock—near Lower Bamham, not far from the Rushgrove Mill point earlier proposed by Green as the canal's southern terminus. (The Duke of Northumberland, whose earlier objections were mainly responsible for the canal being terminated at Druxton under the Act of 1819,[1] which incidentally thus caused the displeased people of Launceston to seek the Tremoutha project, no longer owned Werrington Park, which had been sold in 1864.) The scheme was not proceeded with, however, and instead of a hopefully envisaged extension of trade the canal company suffered an immediate reduction of income, the lowering of tolls on sand and other goods, in efforts to compete with the railway, causing a temporary increase in the tonnage of sand carried but bringing in less money. In August 1878 the committee of management made a straw-clutching resolution that 'the Engineer be requested to report on the feasibility of laying a light tramway from Druxton to Launceston following the River and that he be informed not to communicate with anyone on the matter', but evidently nothing came of this and it proved impossible to retain the captured and draining revenues.

Doom to the canal was further threatened at its other extremity when the London & South Western Railway reached Holsworthy by a branch line from Meldon near Okehampton in 1879. At first this boosted sales of sand which were taken farther inland by rail; following the committee's annual visit of inspection in the summer of 1879 it was reported: 'STANBURY CROSS WHARF—Very good stock of coal and sand.

Man in charge stated that sand is being taken to and over the Railway all day long and that he never remembers the sand trade so brisk as it is now. Messrs Vivian (*the traders*) have fixed a weighbridge on the wharf and a hut adjoining. . . .' But in the 1880s, with farmers turning to the artificial fertilisers and lime now made easily available at the agricultural centres, sand consumption was falling; there was a continuous steady decline too in the tonnage of coal carried by the waterway and a swift drop in that of culm, and figures for tonnages of miscellaneous goods transported and for the harbour and basin dues at Bude showed a similar downward trend.

The situation is illustrated by comparative figures of tonnages and tolls for three periods of four years—the first period before the Launceston railway was opened, the second before the Holsworthy railway was opened, and the third for the four years following:

	Sand Tons	*Coal Tons*	*Culm Tons*	*Goods Tons*	*Tolls (to nearest £)*
Four years ending 1864	158,465	24,070	10,069	10,341	11,929
Four years ending 1876	195,318	16,895	3,128	5,243	7,587
Four years ending 1884	130,719	13,920	4	4,796	5,024

Harbour and basin dues for the same period were:

Four years ending 1864 (to nearest £)	£882
Four years ending 1876	£607
Four years ending 1884	£543

Nevertheless, it was in these later and declining years that the company's only dividends were paid. The first came in 1876, six years after the debt to the Exchequer Bill Loan Commissioners had been paid off and fifty-three years after the canal had opened, and amounted to 10s (50p) per £50 share. Those which followed were: 3s 4d (16½p) for 1877, 5s (25p) each for 1878 and 1879, 6s 8d (33p) for 1880, 3s 4d (16½p) for 1882 and 2s 6d (12½p) for 1899.

CONTINUED MAINTENANCE

Despite the failing fortunes repair and maintenance works were assiduously carried out until well into the 1880s, except that in 1877 the company instructed the engineer to sell the Hobbacott boiler and engine which had become worn out, relying solely thereafter on the well and bucket system for operation of the plane. In 1876 the resident engineer, George Casebourne, had died after over forty years service on the canal; the canal company granted his widow the sum of £50 and also bought her house at Helebridge. Nicholas Sullivan of St Davids, Exeter, was appointed by a secret ballot of the committee of management in Casebourne's place, at a salary of £12 10s (£12.50) per month; the selection was evidently sound and the visiting committee concluded its 1878 report by adding 'that the general state of the Canal and Works is much improved since the appointment of the present Engineer'. (The inspecting party on this occasion included William Pollard, chairman currently of the committee of management, Frederick Pollard —in later years its clerk—and Charles B. Sanders who also at times acted as chairman.) But the need for constant vigilance is apparent from the summing-up paragraph of the 1880 visiting committee's report: 'On the whole the Committee are satisfied that the state of the Canal and Works are improving but Mr Sullivan must be given to understand that he must not relax his efforts, and that particular attention must be paid to the Banks and towing path and culverts and also to cleaning the mud out of the Canal whenever practicable.'

The kinds of jobs which needed attention—dispersed over the canal in much the same pattern as were those which kept the construction inspector and John Honey busy at times earlier in the century—as well as the human aspect, are shown by entries written between January 1873 and November 1882 in a diary kept by James Sleeman, mostly when he was in his thirties. Sleeman worked for the canal company as a carpenter and general handyman, living at Hobbacott, and from his diary one gathers the impression of a man of integrity, perhaps unhurried in his work, yet conscientious, steady and reliable, as were so many country craftsmen of the time.

Entries such as the following tell their own story:

1873	
27 August	Buckett chain broke in North Pit a rivit link.
20 November	Worked in Hobbacottdown Pit today. Water out.
21 November	Chain fell into Pit today. Never saw anything like it before so wild.
1879	
3 January	Marhamchurch Plane altering gear.
1 February	Hobbacottdown sawing gate posts.
23–6 April	Merrifield Plane Putting in Road Timber.
26 May	Hobbacottdown sawing Bridge for Werrington.
18 August	From Holsworthy to Valan (*Vealand*) Plane & to Tamerton Put in Stanks.
29 September	Valan Plane about Pully wheels.
4–8 November	Werrington Plane putting in Main and Pully Beams.
8–12 December	Hobbacottdown sawing and preparing timber for framing at Valan Plane.
1880	
10 January	Reservoir repairs to Bridge and weir.
18–21 May	Reservoir Line General Repairs.
1, 2 & 4 October	Hobbacottdown about Iron for shaft Frame Werrington.
15–17 December	Valan Plane walling under Truss of Water Wheel.
1881	
12 February	Repairing Barrows and to Valan wheel broke.
6–8 April	Bude Repairing Apron to Lock Gate.
28 June	Tamerton Level Repairs Merrifield Plane Chain Broke.
1 September	Blagdon Moor putting in windows to Perkin's Cottage. [*He had been making these during July.*]
29 November	Marhamchurch Plane about segments of Waterwheel.

1882	
14–16 March	Tamerton Plane New small Gear Frame.
17–18 March	Tamerton Plane Arms to Chain Wheel.
20 May	Hobbacottdown shortening Buckett Chain.
November	[*after various repairs—to waterwheel, chain wheel, gear wheel etc the previous month*] Hobbacottdown Plane commenced full duties.

Canal work occupied all but a few of the days of James Sleeman's year. Apart from Sundays, the only official holidays were Good Friday and Christmas Day—and sometimes Old Christmas Day (6 January)—but occasionally he had a day off in summer when he might visit his sisters at Delabole, or go on from work to Bude Fair. Sometimes, however, the duty of

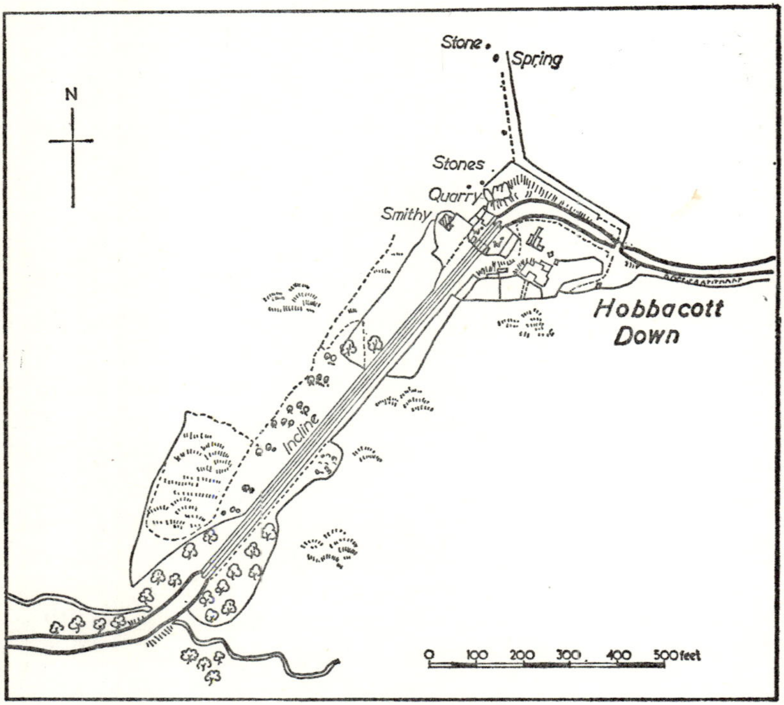

FIGURE 6. Hobbacott inclined plane 1885

attending a funeral or helping a farmer with harvesting took precedence over canal business and there were occasional days —probably when work was slack—when he was 'home jobin'. Sleeman regularly recorded the dates when he tilled and dug his potatoes—in March or April, and September—and he also noted when he killed a pig. His personal courage was called for in those years; in 1879 and 1880 there were a number of such notes as 'home wife very ill', 'home attending wife', 'Nurse Balson with us', 'called home to wife' and then, on 31 August 1880, the sad simple entry:

'Died today my wife about 3 15 am.
Went to Holsworthy.'

After the funeral and the nurse's departure he employed a Mary Jeffery to keep house and must have felt a domestic blank as he resumed his duties on the canal, yet no doubt sustained by his religion for he was a regular attender at Band of Hope functions and, it seems probable, a Sunday School teacher.

DECLINING TRADE AND MOVES TOWARDS INEVITABLE CLOSURE

To offset its expenses, which consisted of salaries, wages and taxes, and maintenance and running costs—always heavy owing to the complicated machinery of the inclined planes (particularly Hobbacott) and the constant repairs necessary to the sea defences at Bude—the canal company received money from several sources. In the twenty years prior to March 1878, as well as the bulk which came in as tolls from traders and other smaller tolls, various lesser amounts were raised by the sale of old metal, chains, faggot wood, hedgewood, bark, canal mud and old stores, from hiring use of a crane at Bude, for the right of fishing and shooting over the reservoir and as contributions from the Lifeboat Association (presumably for its use of canal facilities). George Casebourne, besides being resident engineer, appears also to have been one involved in trading; in 1861 he joined forces with William Dyer for a year and then continued on his own until 1875 when he amalgamated with a Mr Burrow. There was a Thomas Brown operating 1859–65 and Davey & Son 1873–7. By 1877 the only traders on the canal were Vivian & Sons who had started the previous year; from about 1878 the canal company leased them boats, barges and waggons which had been bought following the death of a previous trader, at a yearly rental of £220.

To counter increases in wage rates and the falling trade the company in the 1880s reduced expenses wherever practicable. Workers whose services could be dispensed with were sacked and the salaries of the engineer and clerk were lowered in June 1883 to £125 and £52 per annum respectively. By 1884 the committee of management felt that the expenses had been reduced to the lowest point compatible with safety.

Adding to the gloom were proposals to extend the railway from Holsworthy to Bude. Rumours had been prevalent since 1879 when the company was asked to give its opinion on the scheme (which it declined to do). A few people were of the view that such a railway could be beneficial; when members of the visiting committee journeyed to Holsworthy for the annual inspection in 1879 they were met by Mr Harris of Highampton, who accompanied them some of the way and was strongly of the opinion that the company itself should initiate a project for extension of the railway from Holsworthy to Marhamchurch, which he believed would result in an enormous trade in sand and preclude the railway company's construction of a line right into Bude by another route. In January 1883 it was resolved to prepare and present a petition to Parliament against the railway but this, however, was withdrawn in April when the railway company agreed to insert certain clauses: that the line of the railway would not deviate above a certain amount from the point marked on a deposited plan, that any canal company land used must be by agreement and that any bridges constructed over the canal must clear the towpath by 7ft at least and their design be previously approved by the engineer.

Early in 1884 Vivians announced that they intended giving

FIGURE 7. Scene at the top of the Hobbacott inclined plane during the Canal's working days

up trading at the end of the year. In April the canal company had an overdraft at the bank of £118 5s (£118.25) and had been having to borrow money from the treasurer for current expenses, and a special meeting of the committee was convened to prepare a report.

In the report, which reviewed the position and history of the canal from its commencement and was signed by the committee's chairman, C. B. Sanders, and the secretary, J. F. Eldridge, the following was stated:

> In the opinion of the committee the causes which have led to the decrease of trade are to be found in the fact that lime and artificial manure which are brought into the district by the railways are used by the farmers in the interior instead of sand. The present position and future prospects of agriculture in the district served by the Canal do not favour the hope that sand in any considerable quantity will at remunerative prices ever again be carried over the Canal.
>
> The loss of trade in coal, culm and other goods is also no doubt owing to the competition of the railways, and the Committee are of the opinion that in reference to those articles also there is but little prospect of a revival of trade. The Canal (being mainly made with a view to agricultural purposes) passes through no populous places, and most of its wharves are at a considerable distance from any town.
>
> The diminution in traffic is not transitory, but will most probably be progressive, for in addition to the railways already constructed others have been sanctioned by Parliament, which in the opinion of the Committee will still further prejudice the Company's property; and at present it would appear that when Messrs Vivian & Sons give up the trade at the end of the current year there will be noone to carry on.

The committee felt that the time had come when steps should be taken to wind up the company and realise its assets and it recommended calling a Special Assembly of Proprietors to sanction arrangements for this and to authorise application to Parliament for an Act for the company's dissolution. At the subsequent assembly, held in June that year, the committee's report was adopted and the necessary authority given.

age 87 (*right*)
eorge Casebourne,
ngineer to the Bude
arbour & Canal
ompany from 1832
1876, who super-
sed the rebuilding of
e breakwater after
e disaster in 1838;
elow) A cast-iron
ilepost on the lower
retch of canal near
oddsbridge

Page 88 (*above*) The lock at Roddsbridge. The upper gates have been replaced by a concrete spillway; (*below*) One of the wheels of the tub-boat which was taken from the canal in the autumn of 1970. Although rusty the wheel still turns

On Counsel's advice it was resolved by the committee in September 1884 to apply to Parliament for an Act for the dissolution of the company and the necessary distribution of assets to the shareholders, and the company's solicitors and surveyor were instructed to take the necessary steps for obtaining it in the session 1884–5. There then followed some months of wavering. In October the clerk was instructed to write to Vivians stating the intention of the committee to close the canal at the end of the year unless they would guarantee sufficient traffic to cover the working expenses; it seems that at this point in spite of their earlier decision Vivians were prepared to continue trading, for in November the committee of management made them an offer to carry on in 1885 at existing terms provided they would guarantee a minimum of £1,000 in tolls exclusive of boat hire and basin dues. In the same month an offer was received from a Mr Chamier of Stratton to purchase the Bude Harbour & Canal Company's property for £5,000. Perhaps because of another inquiry received from a Mr Elford, to whom the committee was willing to grant a lease for £2,000 per annum, the Chamier offer was refused at this stage, and Vivians' offer was withdrawn. Within a few days, however, there was another swing-back of the pendulum, the Vivian matter was reopened and consent given to the firm's continued trading for the year 1885 on terms of the existing lease.

Meanwhile the Abandonment Bill had been duly prepared and deposited, had passed Standing Orders and been read for the first time. Before there could be a second reading it was necessary for a special meeting of the proprietors (technically called the 'Wharncliffe Meeting') to be held and the Bill approved by a majority of at least three-quarters of the paid-up capital represented. At the meeting, held on 30 January 1885, the resolution was approved by less than this proportion, and consequently lost. The reason for this contra-decision appeared to be a further offer by a new company under Mr Chamier's formation. The canal company was now prepared to consider selling for not less than £5,000 and wondered whether the prepared Act of Abandonment could be adapted to cover the sale. Further meetings were called and the resolution put that a new company should be formed in time to amend the Act currently before Parliament; however, the new company could not be formed in time for the Bill to proceed in the session and, having already involved the company in costs and expenses of £500–£600, its progress was halted.

During the remaining years of the 1880s the canal managed to sustain a working existence, but with few hopes for the future. By the summer of 1885, due to the uncertainty, Vivians had allowed stocks at the wharves to run low, resulting in prospective purchasers turning increasingly to the railways for their supplies, but in the June prospects brightened when the firm undertook to remit tolls monthly, at which the committee of management decided to keep the canal open at least until the end of December; later an agreement was made with Vivians by which the traders guaranteed payment of a total annual sum of £1,200 in 1886 if the canal was further kept open throughout the year.

At first there was actually an improvement in trading figures, with an increase of 2,786 annual tonnage reported in May 1886. A finance committee was appointed to visit Bude to investigate means of increasing income and reducing expenditure. Regarding expenditure the committee came to the conclusion that there was little possibility of reduction, and considered the amounts paid to employees could not be cut without the proper maintenance and works being endangered. However, there were various means by which it recommended increasing income: by raising charges to firms renting the wharf at Bude for timber, by increasing rates payable by various concerns (including Box's foundry at Marhamchurch) for use of water, and by raising the charge for fishing in the reservoir—currently let to the Rev W. H. Montgomery for trout breeding and shooting rights—from £3 10s (£3.50) to £5 a year. The company resolved to charge 21s (£1.05) each a year for pleasure boats on the canal, while small sums were also raised from the letting of withy beds on the canal route.

Throughout these years, right up to 1890, maintenance works and repairs were diligently attended to. They included making up the towing-paths; repairing culverts; renewing clay puddling; clearing mud from the reservoir and other parts of the system; ditching, fencing and the planting of withies; considerable repairs on the inclined planes including the renewal of various parts; repairs to the locks at Bude and the rebuilding there in 1887 of Falcon bridge, widened at the request of the National Lifeboat Institution. On their annual visits of inspection, which were generally made by boat, members of the visiting committee were continually occupied with disputes regarding the company's property and boundaries; many bound stones, marked with the letters BHC or BC 1839, placed some years previously by the surveyor to mark the boundary of the company's land, had since been removed, and the committee, anxious to maintain its rights, ordered that these should be replaced and further of the boundaries marked, and gave warning of proceedings for any further removals. There were complaints too of cattle straying on to the towing-paths and of farmers tilling adjoining land right down to the water's edge instead of keeping off the regulation 12ft; when this persisted the sub-committee recommended that when dredging the canal the engineer should deposit the mud on the 12ft strip in question in assertion of the company's rights.

Figures presented in May 1887 showed a decrease of 5,691 in the annual tonnage. Vivians having now declined a further minimum guarantee the canal company, desperately aware of the portending inevitability of closure, declared to the public its freedom to arrange with any traders wishing to operate on the canal. At the end of the year, in an effort to compensate for falling trade, tolls on general goods were increased.

By January 1888 it was decided to abandon the canal and to investigate the prospects of Bude either purchasing it or having an arrangement for keeping all or a portion of it open after the abandonment. In response the Stratton Rural Sanitary Authority put forward a scheme for making a reservoir near Halls in Stratton parish but the committee was cautious; what it really had in mind was utilisation of the existing reservoir at Alfardisworthy as the possible source of water supply for Bude. The secretary was instructed to reply that the company objected to the Halls scheme but was prepared to consider furnishing 30,000 gallons daily from the reservoir; he was also instructed to advertise the reservoir fishing in *The Field, Western Morning News* and *Bristol Times*. While the water question simmered boats were being hired to various individuals, farmers began using the canal for drawing sand on their own accounts and Vivians at the end of six months' notice were continuing to trade on a temporary basis. The view was expressed in July 1888 that Messrs Vivian & Sons could very well have complied with requests made by farmers and landowners a few years previously to reduce the price of sand by 4d per ton and still realised a handsome profit; they had declined to do so and the result had proved disastrous to the company.

THE FURTHER POWERS ACT, AND SUBSEQUENT ABANDONMENT

In October 1888 the secretary was instructed to give Parliamentary notices for closing the canal in the current session, but an improvement of business and a profit of £213 10s (£213.50) shown in May 1889 induced the company to defer the ultimate step for a short time. In 1890 tolls were raised, but the annual balance reverted once more to the debit side and the committee resolved to go ahead with the abandonment. Inquiries and discussions about supplying water had continued and after various negotiations with the Stratton Sanitary Authority the committee of management decided to apply to Parliament for an Act for partial abandonment of the canal and to withdraw the (total) Abandonment Act of 1884. The Bude Harbour and Canal (Further Powers) Act (54 & 55 Vict c 75) was passed on 3 July 1891 and authorised abandonment of the branches from

Red Post to Druxton and from Brendon Moor junction to Blagdonmoor, the maintenance of the harbour at Bude and of other portions of the canal and the reservoir. The towns of Stratton and Bude, which had hitherto been dependent on unsatisfactory wells and pumps, were to be provided with a regular and continuous flow of water brought from the reservoir along the canal to the top of Hobbacott plane, for which the local authority was to pay the canal company 1d per 1,000 gallons up to 60,000 gallons a day and ½d per 1,000 gallons taken in excess, the local authority to construct and maintain the necessary sluices, filter beds and piping. So that it could suitably maintain the portion of the canal from the reservoir to Bude the company was empowered by the Act to raise either by shares or borrowing a sum not exceeding £10,000.

The Act also sanctioned an arrangement which had been made by Edward Mucklow, JP, of Whitstone for the supply of water from the canal company for property belonging to him in the parishes of Marhamchurch and Poundstock. Edward Mucklow had become an increasingly prominent and influential member of the canal company in the declining years of its life, during which he acquired controlling powers and was a notable and colourful figure in the neighbourhood. Thick-set, heavily-featured and bearded, he had come to Whitstone from the north during the latter part of the nineteenth century and was soon accredited locally with a fantastic reputation of immense wealth. A particular theme seemed to recur in the stories about him which went the rounds: an old lady would recall how, when he arrived in the district, he brought with him 'a waggon loaded with boxes of gold' and it was also said he would give 'a cartload of gold' to anyone who would build a railway to Bude. Although such tales are doubtless exaggerations, when he came he was certainly extremely wealthy, but it was also said that he lost over a million pounds in the district, largely due to his eccentricities. He bought a number of farms in the area on which he straightway erected new farmhouses and buildings, and built Whitstone Head for himself; extravagant gossip claimed that, so particular was he about the quality of work, if he hadn't seen the actual laying of a house's foundations he would have them removed and relaid under his eyes. He 'would never use wood if stone would do', so it was told.

With much of his land lying in proximity to the canal and being himself an instigator of the extension of the railway to Bude, Mucklow was probably at least partly influenced by his own interests in the work he did for the canal company. Held in awe by some, feared or respected by others for his financial wizardry, obsequiously thanked on occasion for genuine help which he gave, he could at times be stubborn and infuriating to fellow shareholders. He and his two sons, Walter and Edward Jnr, all sat as members of the committee of management whose meetings he chaired on numerous occasions, and he was also the last chairman of the company.

Immediately after the passing of the Further Powers Act the committee of management began making the necessary moves to close the Launceston and Holsworthy branches and to realise the company's property along these lines. Notice of the intended closure of the two branches on 14 November 1891 was given and a week's notice to quit issued to most of the employees not already dismissed, excluding the harbourmaster, Thomas Found. It was resolved to give a month's notice to Mr Davey, clerk since 1863, and to the engineer, Mr Sullivan, but on consideration both were retained at reduced salaries. (William Davey retired later, in 1899, and Sullivan resigned in 1895, continuing thereafter as consulting engineer to the company.) Sales of old materials and equipment were held, at Druxton by Mr Kittow on 14 September, and at Helebridge on 26 October when Mr Sandry was the auctioneer, and other property was disposed of. Some boats were sold and the breaking up of others commenced, and old chains, scrap iron, timber and stone were collected up and traded. The engineer was ordered to secure against accident the two wells at Hobbacott and to keep down wages to the remaining employees. Landowners were offered the option of purchasing canal land adjoining their own at a reasonable price and early in February 1892 Mr Thomas Shearm of Bennetts, Holsworthy, was engaged as valuer on the company's behalf, at a fee of a guinea (£1.05) a day, including travelling expenses.

The Realisation of Property Account presented to the committee of management, dated 30 June 1892, showed a financial return of nearly a thousand pounds after deduction of expenses:

To Lands & Buildings realised	635	0	0	By Fee for valuation of Hele Bridge property	2	2	0
To Iron, Timber, Stone & Boats realised	533	7	11	By Day Labour (Breaking up Planes & taking Material to places of Sale etc)	195	14	10
				Balance Surplus	970	11	1
	£1168	7	11		£1168	7	11

PROTRACTED NEGOTIATIONS AND EVENTUAL SALE

Difficulties had arisen, however, concerning the prospective supply of water to Bude and district, due to the Local Government Board having refused to give sanction to the loan required by the Sanitary Authority for carrying out the scheme. At its August meeting the committee held out no hope of working any remaining part of the canal at a profit and feared that the shareholders would have to prepare to abandon finally the whole undertaking. The following month, when the visiting committee walked over the whole of the unabandoned portion of the canal from the head of the reservoir to Bude, it was apparent that any supply of water taken from the open canal at Hobbacott would be scarcely fit for domestic purposes, and at a subsequent meeting between the committee and members of the Bude Sanitary Authority suggestions were made for taking the water direct from the reservoir instead of from Hobbacott.

After consideration the committee decided it would be prepared to supply water from this point on the same terms as for the previous arrangement.

But the unsatisfactory and unresolved situation dragged on through 1893 to 1894. In each of these years a loss was reported in the revenue account and although the sale of land on the abandoned portions of the canal progressed steadily until completion—realising £2,284 5s 7d (£2,284.28)—in June 1894, the committee was quite unable to come to any terms over the water supply. The committee of management's report to the meeting of proprietors held on 23 February 1894, signed by Henry Tapscott as chairman and F. Pollard, secretary, contained the following:

> The Committee understand that the Bude and Stratton Sanitary Authority have now definitely give up any intention of obtaining a water supply for their district from the Canal or Reservoir, and the Committee cannot hold out any hopes that any scheme for such supply will be carried into effect.
>
> It does not appear likely that the future receipts from the Harbour Dues and Canal Dues from Bude to Marhamchurch will suffice to pay the working expenses and the Committee consider it probable that the entire undertaking will have to be abandoned, and the affairs of the Company wound up at no distant date.

In March 1895 the committee approached the London & South Western Railway Company with a view to being taken over, but the railway company replied that it had no intention of purchasing the canal company's property. It was then resolved that if freed from all liability the company would accept £2,000 net (raised to £2,500 the following year) for its entire rights and assets, the purchasers to obtain an Act of Parliament and pay all expenses connected with the sale, and Mr Mucklow was authorised and requested to endeavour to find a purchaser. Negotiations for the sale were reported to be in progress later that year and in the next. In November 1896, following apparent inquiries from the newly-formed Stratton Rural District Council, the committee replied that the company would not enter into negotiations for supplying water but was prepared to negotiate with the council for the absolute sale of the reservoir and water rights.

A complication in the offer of the canal for sale lay in the scheduled agreements in the Further Powers Act of 1891 by which Edward Mucklow was allowed to abstract from the canal for watering his property. This led to a stormy meeting of the committee of management on 11 December 1896 when, with Mucklow in the chair, it was reported by C. T. K. Roberts, the company's solicitor, that there was little prospect of selling the canal until these agreements were modified to meet the view of Mucklow and the Stratton RDC. A resolution was duly moved and seconded only to be torn up by Chairman Mucklow, who refused to put it to the meeting and left the room. In the stir which followed Colonel C. Halford Thompson of Teignmouth was voted to the chair and in the former chairman's absence it was resolved that as Mucklow was not willing to waive his rights under the Act the company was not in a position to negotiate for the water rights with the Stratton RDC. Colonel Halford Thompson was so disgusted with Mucklow's conduct at the meeting and the way in which he was using his controlling power, prejudicial to the interest of the proprietors as a whole, that he resigned his seat on the committee forthwith.

Early in 1897 a firm of Dickson & Company showed some interest in the canal, and asked the company for an option to purchase at £5,500 at any time within three months, but the request was not pursued. Later the same year Edward Mucklow gave news of a syndicate of which he was one, which desired to have an option for one month to purchase the undertaking and assets of the company at £3,500; his intention was the formation of a new company which would have the right of supplying the district with water and would be willing also to take over the responsibilities regarding maintenance of the breakwater, harbour and basins at Bude and the canal's lower reaches, and also itself to finance the necessary Act. The matter was discussed at meetings of the committee and the offer accepted providing the sum was raised to £3,622,[2] but Mucklow wrote soon after that he could not make any progress in the formation of a syndicate, and the project was dropped.

Meanwhile negotiations were proceeding for purchase of certain lands and easements required by the railway company, whose impending line from Holsworthy to Bude was expected to bring material increases in harbour and basin dues. There was some bother over construction of the railway arches across the canal at Helebridge which caused a narrowing of the towpath and an accumulation of silt due to diversion of the River Strat below the bridge. And when two barges which contained cement belonging to Aird & Company, contractors for the railway company, sank at Roddsbridge, the canal company was sent a bill for £16, the value of the cement, by the resident engineer Mr Grainger; liability, however, was not accepted. The railway was opened in August 1898 and although there was a small amount of traffic from ship to rail, harbour and basin dues showed a definite drop, from £214 in 1897 to £173 in 1898, £134 in 1899 and £125 in 1900, while trade on the canal itself apparently ceased altogether. Small sums of money were realised by the canal company in these last days from the sale of withies for basketmaking, from water sold in small quantities and to the railway company, from bill-posting space let at Bude and the sale of fishing permits and from rents for land and buildings.

During 1898 there were further moves towards selling the canal. In October that year a suggested agreement was drawn up by which the local authority would take over the reservoir and canal to Hobbacott, constructing necessary waterworks and bearing two-thirds of the capital cost, with Edward Mucklow finding the remaining one-third and in return having the right to take a third of the available supply. Working expenses and 3 per cent interest on the works' capital cost were to be deducted from the sum payable to the canal company for the water. But the committee resolved not to enter into any such negotiation with the local authority unless it could first be arranged that the originally agreed price of the water supply should be substantially increased.

By the end of the year it was again decided to try and sell the whole of the remaining property and undertaking and early in 1899 the Stratton District Council was informed that the committee would recommend acceptance by the shareholders of the sum of £8,000 for the whole of the assets (other than money in hand and debts due). As an alternative the company was prepared to sell water to the council *from the reservoir* at the rate of 3d per 1,000 gallons up to 10 million gallons, 2d per 1,000 for the next 10 million and 1d for any surplus. The Stratton District Council unanimously adopted the recommendations but there was shortly a hitch due to the Local Government Board declining to sanction purchase of the canal by the council for the sum asked.

In 1900, however, further negotiations proceeded between sub-committees of the Bude Harbour & Canal Company and the Stratton and Bude Urban District Council, and the council subsequently resolved to promote a Bill to Parliament for the purpose of purchasing the entire undertaking at the previously-asked figure of £8,000. Meanwhile, and in the event of failure to obtain an Act, the company agreed to supply water to the council at Vealand plane.

After preparation of the necessary agreement by the company's solicitors it was stated that Edward Mucklow's solicitor had made alterations in it which would put unacceptable obligations on both the company and the UDC. Presumably these concerned Mucklow's rights to the water which would be by-passed with the taking of the water at Vealand; the UDC was prepared to proceed, however, on the undertaking that purchase or lease should be subject to Mucklow's rights under the Act of 1891.

Following a somewhat precarious passage through Parliament the Stratton and Bude Improvement Act was passed, and received the royal assent on 17 April 1901 (1 Edward VII c 258).

It was expected that the profit of the canal, stated to be £6 for the previous half-year, could be maintained and probably increased, and it was estimated that the annual revenue for water would be £500. Powers were given for borrowing money needed for the purchase and for financing the various necessary works.

No. 67

Bude Harbour and Canal Company.

1, Bedford Circus, Exeter,
5th March, 1902.

Dear Sir or Madam,

I have the pleasure to inform you that at the General Meeting of Proprietors, held on February 14th, 1902, it was resolved that a sum of Four Pounds per Share be returned to the Shareholders; and herewith I send you a Cheque for £4 the amount payable to you in respect of the one Shares standing in your name in the Company's books.

The Receipt at the foot of the Cheque must be stamped, signed, and dated by you, and the Cheque then presented for payment through a banker in the usual way. No further acknowledgment is necessary.

Your obedient Servant,
FRED POLLARD,
Clerk.

James Higgs Esq

Figure 8. Communication of final dividend 1902

Sold for one-fifteenth of its original cost, the canal was formally handed over to the council on 1 January 1902, and two months later sums amounting to £4 per share were paid out to the various shareholders. Having fallen victim, like other canals in the westcountry and elsewhere, to the development of steam locomotion and the railway system, the Bude Canal as an inland waterway had reached its eventual but inevitable demise.

CHAPTER 5

The Twentieth Century Scene

SOON after acquiring ownership of the canal's remaining length—the main trunk (minus now the Holsworthy and Launceston branches) and the feeder arm and reservoir—the Stratton & Bude Urban District Council began construction of its new waterworks. The existing 4¾ mile stretch of canal leading from the reservoir was retained as an open aqueduct and the water extracted at Venn, Pancrasweek. Here, near the top of the Vealand inclined plane, filter beds were made as authorised by the Act of 1901 (instead of at Hobbacott as had been envisaged in 1891) and piping was laid to carry the water after treatment to the urban area. The council permitted fishing to continue in the reservoir, which became known as Tamar Lake, and provided further encouragement by stocking it with brown and rainbow trout. The lake over the years has become established as a sanctuary for many birds which in 1950 gained official protection by the passing of an Order which prohibited the destroying or taking in the area of wild birds or their eggs.

For many years the length of canal from Vealand down to Helebridge lay idle and virtually untouched, yearly becoming more derelict. For the most part the towing-path remained passable and the council did not exercise its powers officially to exclude the public as it did on the water-carrying section from Venn to Tamar Lake. From Helebridge to the harbour stopping-up of the towpath was not allowed under the Act and this section was actively maintained for both port and amenity purposes. This at times involved the council, as harbour authority, in considerable expense. Early in the century a terrific storm in 1903 and a phenomenal tidal wave the following year caused much damage; repairs were needed to the breakwater, and new lock gates had to be provided following the tidal wave, which entirely destroyed the former gates, sweeping them out to sea, and released to waste a large volume of water from the canal. There was dwindling trade with sea-going ships, which continued to enter the canal to berth at the quay on which a railway siding had been built. The edge rail tramway was used to carry sand from the beach to this lower wharf where it ran up a slope enabling the sand to be tipped directly into the railway trucks; this traffic ceased in the 1940s but the siding continued to be used to bring coal and other goods to W. W. Petherick's store until 1964, two years before closure of the railway in 1966. Pleasure boating was permitted, and still continues, on the canal's lower section and access by the public is freely allowed along the 2 mile towing-path.

By the 1950s the council was facing various problems concerning the canal. Under the law the whole 2 miles of canal from the sea lock to the bottom of the Marhamchurch inclined plane was still officially a navigable waterway which the council was under obligation to maintain. In practice, however, navigation of the full stretch was impossible as the locks at Rodds-bridge and Whalesborough had been concreted since the 1920s and could not be opened, and there had been considerable silting. One thorny problem concerned Falcon bridge, situated between the lower and upper basins, which had become inadequate and dangerous for modern traffic. By the original Act of 1819 the canal's owners were required to maintain it as a swing bridge to permit the passage of shipping (though nearly forty years had passed since it was last opened and in fact the bridge had been secured by the council for some time), and the Cornwall County Council was not prepared to consider building a new bridge to carry the highway until the legal position was changed. The 6½ mile stretch of abandoned and overgrown canal from Marhamchurch to Vealand was another matter for concern as it was alleged that damage from water and other causes was occurring to adjoining land due to neglect and improper maintenance. At the same time the cost of maintaining the obsolete Burmsdon aqueduct over the Tamar had necessitated the council borrowing money for repairs.

Deliberations led to the presenting of a Bill to Parliament and the passing in 1960 of the Bude–Stratton Urban District Council Act (8 & 9 Eliz II c 28). By this the council (re-titled the Bude–Stratton UDC in 1934) was empowered to dispose of disused canal land from Helebridge to Vealand (except for Burmsdon bridge and the Tamar aqueduct), rights of navigation between Roddsbridge lock and Marhamchurch were extinguished (thus legalising stopping of the locks) and the requirement that Falcon bridge should be capable of being opened was removed.

Within the next year or so sections of the disused length of canal from Helebridge to Vealand were sold, most of them, like the greater part of the former Holsworthy and Launceston branches, being incorporated back into farm land. In 1967 Tamar Lake and the canal from it as far as Burmsdon were transferred to the North Devon Water Board. Apart from one portion of the disused length—a piece near Anderton which the farmer concerned has not purchased—the only part of the canal remaining in the ownership of the Bude–Stratton UDC is the 2 mile length from Bude.

The Present State of the Canal Remains

Although described here in some detail, it must be emphasised that by far the greater part of the former canal's route lies through private land and is therefore not freely accessible by the public. There are in fact no public rights of way over the route, although the Bude–Stratton UDC allows (and encourages) members of the public to walk along the section of the towing path from Bude which remains its property, and does not object to the public walking over the disused stretch near Anderton which it still owns. There are, however, many access points on public roads from which the line of the canal can be determined and where buildings and other features are easily observed. The features as described are as they appeared during the authors' recent investigations and do not take into account any further deterioration or changes which may have occurred in the immediate pre-publication period.

SEA LOCK TO HELEBRIDGE

Standing beside the sea lock today, it is difficult to imagine the canal as it was a hundred years ago, when the basin would have been full of sailing ships discharging their cargoes and awaiting the tide to enable them to return to sea. Looking out to the west the breakwater can be seen stretching from a few hundred yards north of Compass Point to Chapel Rock, then on past Sir Thomas's bathing pool to the Barrel Rock, on which a barrel is still fixed as a landmark for incoming boats. The course of the channel which had to be followed by craft approaching and leaving the canal is still marked by mooring posts and buoys, but, in spite of these and the breakwater, passage must always have been hazardous under sail and only possible with a certain depth of water. Climbing over the rocks past Whale Rock towards Compass Point it is just possible to discern the position of the railway which was laid in 1838 to enable rocks to be brought from Compass Cove for rebuilding the breakwater after the great storm; remains of iron posts embedded in the rock with their associated rust stains are clues to its path.

Turning back inland, the sea lock with its massively constructed gates is seen prominently marking the canal's entrance. Here, and around the basins, there are numerous remains of the canal's commercial days. On both sides of the sea lock basin there are a number of mooring rings. On the northern side of the canal a slope from the basin edge leads down to the beach, crossing a bridge over the River Strat, down which the edge railway was built to enable truckloads of sand to be dragged up by horses and tipped into the tub-boats and barges moored beside the basin. In about 1910 the original 4ft gauge lines were replaced with rails of 2ft gauge; remains of the latter can still be found and there are some granite slabs with holes bored in them showing how the rails were anchored. Sometimes the shifting sands expose a turntable at the foot of the bridge, and formerly there was an extensive network of rails stretching out across the bay. On the north side of the canal under Castle Hill are some early canal buildings; the first group comprise the forge and harbourmaster's office and then, where the basin widens, set back a little is the one-storeyed stone building with a louvred slate roof, which has been used variously as a salt store, tannery and a warehouse for lime. Other former warehouse buildings are now used for storing coal and other purposes.

The present Falcon bridge replaces a swivel bridge which opened to allow ships to proceed from the lower wharf to the upper one, which was known as Sir Thomas Acland's wharf. Over the bridge on the west side of the canal are a number of contemporary buildings. First is the lifeboat house presented in June 1865 in memory of Elizabeth Moore Garden; the launching of the lifeboat from this building was a source of some trouble to the canal company due to damage to Falcon bridge which its passage caused. A large three-storey slate-roofed building with a projecting wing was one of the first to be built on the upper wharf, next to it is a long white-painted store house and above them was once a flourishing shipbuilding yard and sawmill. Farther along a former steam laundry building still stands, now used as a store by the council, and there are remains of limekilns and a little cottage which was once known as Round's Casting House. Above the cottage, at the foot of a private garden, is a lay-by or turning place for the barges.

The waters of the first stretch of canal, often used for pleasure boating, probably conceal many sunken relics of its working days. In 1970 the council recovered a long-submerged tub-boat which it plans to preserve and in 1971 another tub-boat was retrieved by members of the Exeter Maritime Museum where it is destined to form part of an exhibit. There is a towing-path here on the canal's east side as far as the road bridge which crosses at Roddsbridge Farm; just before this is reached a cast iron milepost marked Bude 1 can be seen, and at the bridge the towing path passes to the canal's west side.

Within a short distance are the remains of Roddsbridge lock, and a little farther along the one at Whalesborough. Both locks have been stripped of their gates and mechanism—though the positions of the gates can still be seen—and in each case the upper gates have been replaced by a concrete spillway.

Above here the canal is overgrown. In a few hundred yards there is an occupation bridge with the original stone footing, and just past it, on the opposite side of the canal from the towing-path, is the weir over which the River Strat flows on its way to the sea. Presently the path crosses a tributary of the river and immediately ahead is the now derelict railway arch which crosses both road and waterway. Emerging from under the railway bridge the canal runs beneath the original Helebridge which carries the road over the canal and the River Strat. Incorporated in the wall between the two bridge sections is one of the old bound stones of the canal company, marked BC 1839.

HELEBRIDGE TO VENN

Helebridge was the terminal for the barges from which goods were unloaded. Sand and coal in particular were stored here for direct sale but most of the goods were transferred to tub-boats for passage up the incline to the next level of canal. Standing against the road bridge on the south side of the canal is a small two-storey building, recently modernised, which at one time was used for barge repairs and sailmaking, with a sawpit situated actually inside. The full extent of the basin is difficult to determine as it is very overgrown, but on the southern bank at the east end of the basin stands the remains of a warehouse, and the massive stone edge to the wharf can be traced. Between the warehouse and the barge repair shop is the house which was inhabited by George Casebourne, the engineer from 1832–76. A pair of white cottages standing on the north side of the road were formerly stables for the horses which towed the boats.

The towpath follows the north side of the basin to the commencement of the Marhamchurch inclined plane, which is now completely overgrown. Near the foot of the incline, on the same side, are the remains of Box's iron foundry from which a track leads up beside the incline to Marhamchurch village above. The pit at the top which housed the 50ft waterwheel that worked the plane's mechanism has only recently been destroyed and is now completely buried.

From the top of the incline the route of the canal, crossed by the track from the foundry, continues through fields and alongside gardens, and passes beneath the Marhamchurch–Stratton road. It is said that a former rector here had a private bridge across the canal between the rectory and the church. Following the contours the canal route turns northwards towards Cann Orchard, crossing a stream which runs through a well-constructed culvert; before reaching the farm there are remains of three occupation bridges, but two road bridges in its vicinity have been completely obliterated.

Beyond Cann Orchard a considerable length of the canal has been ploughed in. There are the remains of another occupation bridge and as the line approaches the foot of the Hobbacott incline it is raised on an embankment running through a copse, and turns east over a very fine culvert through which flows a tributary of the River Neet. At the base of the incline are two well-defined channels which indicate where the tub-boats waited before being hitched on to the endless chain which drew them up the plane when the mechanism was set in motion.

The inclined plane is impressive even in its present state. There is still a granite block to be found with a central hole, which may indicate the position of the line of rail. Half-way up, on the right, is the site of the entrance to what may have been an inspection pit, but of the pit there is no trace. At the top the keeper's cottage still stands, but is steadily becoming more derelict and there is another building whose former purpose is not known. One of the two wells down which the buckets descended is visible behind an iron grating, stone-lined and in a remarkable state of preservation; the other pit has been covered in. Behind the buildings the position of the lay-by, where the boats would have collected to await their turn to continue their journey, can be discerned. One can imagine the scene of activity here during the canal's working days; an elderly lady living across the valley remembers hearing, as a child, the clank of machinery when the boats were being pulled up the plane.

From Hobbacott Down to Red Post the canal has been ploughed in and the only visible remains are of the abutments of the aqueduct which carried it over the road to Thurlibeer Farm. The route crosses the Launceston–Kilkhampton road (B3254) just south of the Red Post crossroads, a tell-tale hump in the road marking the spot where the canal passed beneath Thorn bridge, and here, close to the site of an old brick and tile works, the Launceston branch leaves the main Bude–Holsworthy line.

The Holsworthy line, continuing past the old brick works, is difficult to follow owing to undergrowth, the track alongside being composed even now mostly of brick rubble. The course can be picked out again as it skirts a field east of the Shernick Farm lane, where there must have been an occupation bridge, though no traces are apparent. For about half a mile northwards there is a well-marked towing-path on the canal's east side, with signs of a lay-by and remains of the abutments of an occupation bridge. Curving with the contours the route passes behind a modern bungalow before crossing the main Holsworthy–Bude road (A3072) with only a faint hump in the road showing its position.

North of the road the canal route plunges into the undergrowth and is built up on an embankment through which there is a culvert for a tributary of the River Tamar. It emerges from the woodland and runs along the edge of the field; there is a good towpath and the canal bed is obvious with reeds and irises marking its course. A new concrete bridge over the former canal gives access to Anderton Farm, the towpath continues and about ¾ mile on there are the footings of an old bridge. The towpath is sometimes overgrown along the next stretch but the canal's course is always obvious, sometimes dried out and sometimes wet and filled with rushes and other damp-loving plants.

The route now approaches what used to be called 'the long turn' at Burmsdon. Here it is again raised on an embankment with a culvert beneath for a tributary of the Tamar. The towpath is clearly apparent and the canal keeps to the contour until it reaches Burmsdon Farm bridge, which is a perfect example of a typical canal bridge with the towpath running under it. The line continues below Burmsdon Farm to the fine stone-arched aqueduct which carried the canal over the River Tamar; this now carries the pipes of the water supply from the treatment works and in recent years has lost some of its former beauty through the removal of its parapets and the erection of a barbed wire fence in their place. The canal route runs eastwards on an embankment, with a culvert beneath carrying the leat for Kingford Mill, and a little farther on the remains of an occupation bridge. Here the canal is filled with vegetation but its site is obvious. A slight right turn indicates the beginning of the Vealand inclined plane, though there is nothing to show where the tub-boats waited nor any indication at the top as to the position of the waterwheel which provided the power. A small modern brick building at the foot of the plane is owned by the North Devon Water Board and used for water chlorinating; the filter beds are at the top.

VENN TO HOLSWORTHY

Along the next $\frac{1}{2}$ mile from the treatment works the canal continues in a north-easterly direction and is full of water; there is a good towpath and, apart from a new bridge leading to Vealand Farm, the scene must resemble the general appearance of the canal in its working days. At the Brendon Moor junction, just below Brendon bridge, the surviving character continues only in the feeder arm which branches north; all that can be seen of the Holsworthy line is a slight constriction in the channel bearing away east and then south-east; the route here is very overgrown but is discernible from the road at Lishaperhill and again at Vicarage (now Glebe) Cottage. It is built up on an embankment above the level of the surrounding fields, crosses a valley over Small Brook and runs through some woods to pass beneath a bridge carrying the Thornemoor road, now a dumping place for rubbish. The next section is impassable but the route is traceable again in the woods where there are definite signs of a lay-by or passing place.

A roving or footbridge which still exists where the footpath from Burnard's House used to cross the canal has iron girders similar to those seen in bridges over the feeder arm. The route, virtually impenetrable, then passes through Parnacott Deep Cutting to the bridge carrying the road from Burnard's House, where it has again become used as a convenient rubbish dump, and, continuing north-eastwards, runs through woods and across a field to the first of two roads to Parnacott Farm, where there was once a wharf. Signs of the wharf, and also of the bridge through which the canal passed beneath the road are gone. Along the next stretch, across fields, the towpath is prominent, the canal itself having in some places been ploughed in and in others left uncultivated and full of reeds; at the crossing of the second Parnacott Farm road there is no indication of the bridge which must formerly have existed.

The course turns northwards and in the woods beside Canal Farm there are still some of the boundary stones put down in

1839. There is a good culvert under the canal through which a stream flows and from here to the main Holsworthy–Chilsworthy road the farm track has been made along the line of the canal. Only recently has infilling and levelling eliminated the bridge by which the canal passed beneath the road. The course of the canal continues beside a private road to the Chilsworthy embankment which is very impressive with a large tunnel-like culvert through it in which flows the River Deer; at the embankment's western end stables used in the past for canal horses still stand together with a cottage which probably belonged to the canal company.

From east of the embankment the canal route passes through fields at Hogspark; here it has been ploughed in, though the position of a lay-by is indicated by an area where irises grow. Some imagination is required to trace the continuing course to the high Manworthy embankment; also containing a culvert, this embankment still carries the towpath although the canal bed appears to have been cut away to enlarge the field. On its west side is one of the larger bound stones, BC 1839. After the embankment the course is apparent through the fields; the towpath is in places clear and in others overgrown, but the bed of the canal is damp and reedy and easily recognisable.

Meandering south and then eastwards, the route crosses and runs alongside minor roads and then curves away across fields to Stanbury wharf, beside the Holsworthy–Bideford road (A388), where there is the Bude 15 milepost. The old stone-built shed where goods were stored still stands, its particular style of construction with rather low walls and broad-pitched roof being similar to warehouses built by the canal company at various other wharves. It is now converted for use as a cow shippon, dairy and haystore and the openings through which materials from the boats were unloaded have been closed over. The wharfinger's cottage still stands and is inhabited and there is also in existence a pair of anchor stones with chains attached.

The canal passed beneath the now-level road by a small bridge supported by iron girders for the final mile across fields

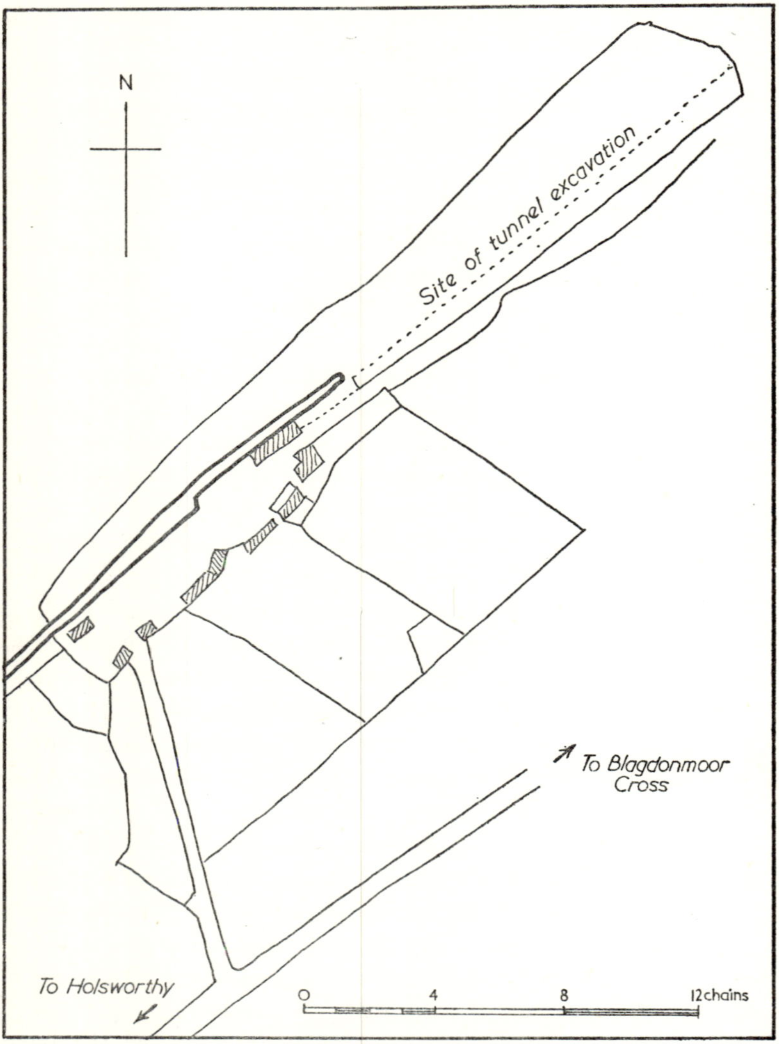

FIGURE 9. Blagdonmoor wharf 1843

to Blagdonmoor wharf. The basin here, situated at the end of a short lane from the road to Thornbury, is now choked with weeds but several buildings remain, including a warehouse like the one at Stanbury which was used for coal, sand and other goods, stables and cottages for the employees. It is of interest that, although traffic terminated here, the excavation of the canal continues past the last warehouse and enters a considerable cutting which marks the work on the intended tunnel on the projected Thornbury line.

THE FEEDER ARM AND RESERVOIR

From Brendon Moor the feeder arm, bringing water to the treatment works, covers a meandering course northwards to the reservoir. As already remarked, the character of the water-filled length must give a good impression of the appearance of the canal in former days and on this stretch there are a number of small road bridges crossing the waterway from which it can be easily observed. First of these above the junction is Brendon bridge, a new road bridge built on the old footings, which has slots on its north side into which planks could be fitted as a stop lock during repair work. Then, after three similar occupation bridges on the next stretch, there is Puckland bridge, a rebuilt brick arched bridge again constructed on the old footings. Farther on is Moreton bridge which has been widened on each side and has wooden rails and is unusual in still having the original four cast iron arches in position; next is Dexbeer bridge, a modern brick structure with a very low arch.

There are three further occupation bridges on the next section which includes a raised embankment under which a small stream flows through a culvert; one of them, like Moreton bridge, still retaining the cast iron arches. Another embankment carries the canal over a valley by a very sharp U-turn (known previously as Cape Horn) and the towpath ends at a modern bridge over the canal at the first Virworthy wharf. Here there are typical store buildings used in the past for coal, sand and manures and now for farm purposes, and there was also a mill which still stands though the millpond is filled in. The navigable part of the canal ended just a short distance farther north at another wharf beside some cottages. These have recently been converted into one house and the owner found a considerable quantity of coal and sand when digging the garden over the site of the wharf.

The 70 acre reservoir, Tamar Lake, today presents a scene of considerable mellowed beauty and is of great interest to ornithologists and anglers. It is roughly triangular in shape, with a sluice at the northern end for regulating the amount of water passing into the reservoir; the required compensation flow, of particular importance in the past for mills which existed downstream, running along the river's altered course on the west side. The dam, sited across the reservoir's southern end, is said to be steadily sinking and is considered unsafe and plans have been put forward for a replacement reservoir on an alternative site. Destruction of the dam, either by accident or design, and the resulting draining away of the lake, which in this particular setting is such an asset, would be a sad loss from both the amenity and conservation aspects.

THE LAUNCESTON BRANCH

The route of the Launceston branch of the canal leaves the main Holsworthy line just east of Thorn bridge at Red Post. The first few hundred yards have been almost completely obliterated by ploughing, but then it is raised on an embankment for a short distance before going into a cutting beneath the Treyeo farm road where it is being filled with rubbish. In Burn Wood the canal has been partially obliterated although the towpath is clear; the line follows the contours of the hillside, and when the Grove farm road is reached the bed is easy to discern with a quarry on the west side. The next section, though now completely ploughed in, can be made out at a distance, the line of irises, sand and clay giving away the course. The culvert over

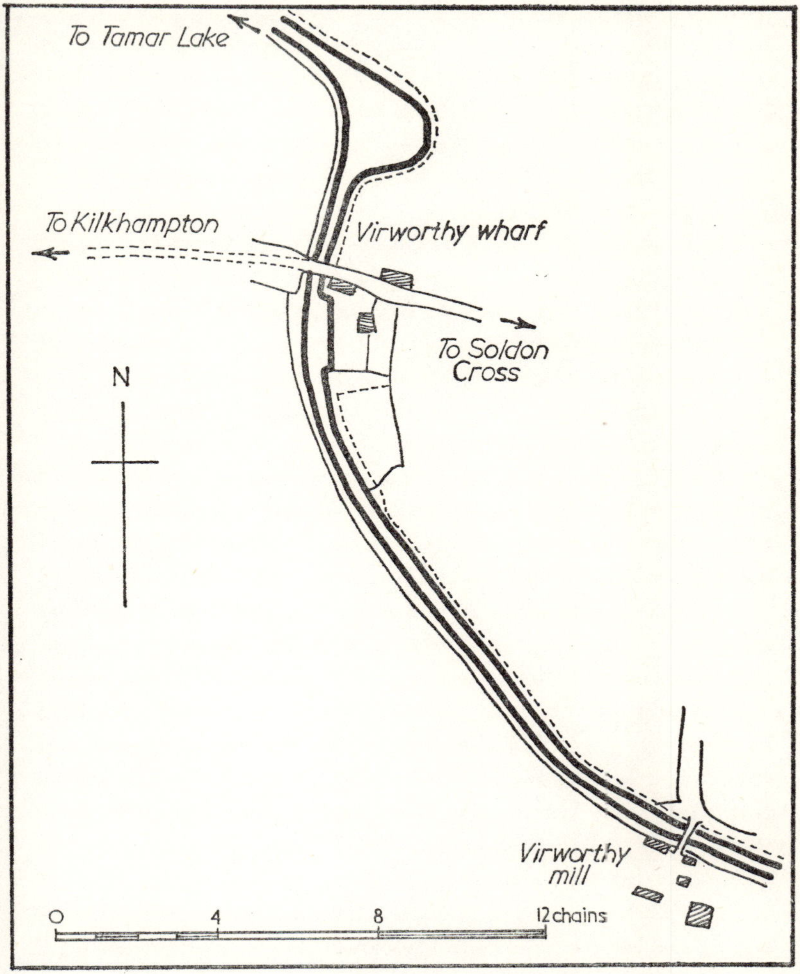

FIGURE 10. Virworthy wharf 1890

a stream which flows into the Tamar has been broken up and the canal route, following the contours in its approach to Littlebridge, Bridgerule, and partly in a cutting, is very overgrown.

At Littlebridge, on the east side of the road to Moorside, there was a wharf, and an elderly lady living nearby can remember boats lying here derelict after the canal closed. The wharf later became a sawmill and the former wharfinger's cob-built cottage, which stood on the right as one entered the wharf from the road, was demolished. The canal crossed the road by a high aqueduct, the supporting pillars of which can still be seen, and there is an anchor stone reposing by the roadside a short distance away.

The canal swung south just east of Borough Cross and passed beneath the road to Bridgerule. Apart from a short raised length the next section is hard to determine due to levelling for agriculture, but the damp, iris-growing depression east of Newacott is evidence of the former lay-by there.

At Merrifield the first of the Launceston branch's three descending inclined planes is overgrown but still recognisable. At the top of the slope, on the eastern side, is the stone-arched entrance to the wheelpit, which is about 28ft in length internally and of 9ft 6in breadth. On three sides of the pit there is a ledge roughly 2ft wide which enabled the incline keeper or his assistant to have access to the wheel, and also to a chamber built on the side opposite from the entrance, where the machinery which set in motion the chain for raising and lowering the tub-boats was apparently housed. There are the remains of a hatch in the roof over the wheelpit itself, and another in the roof of the side chamber. At the bottom of the pit on the east side is an arched passageway which apparently carried away the waste water; it is 32ft long but the end is now blocked up and the exit on the hillside is only conjectural.

The railway having been built after the abandonment of this part of the canal, the continuing canal route is severed abruptly at the now derelict railway and continues on the farther side. For the next stretch the canal bed has been ploughed in and only the towpath is obvious, the canal route, the Langaton Mill leat and the River Tamar in places running in very close proximity. The canal route passes through a shallow cutting as it approaches the Langaton Mill road, south of which the towpath and canal bed are well marked for a spell until entering

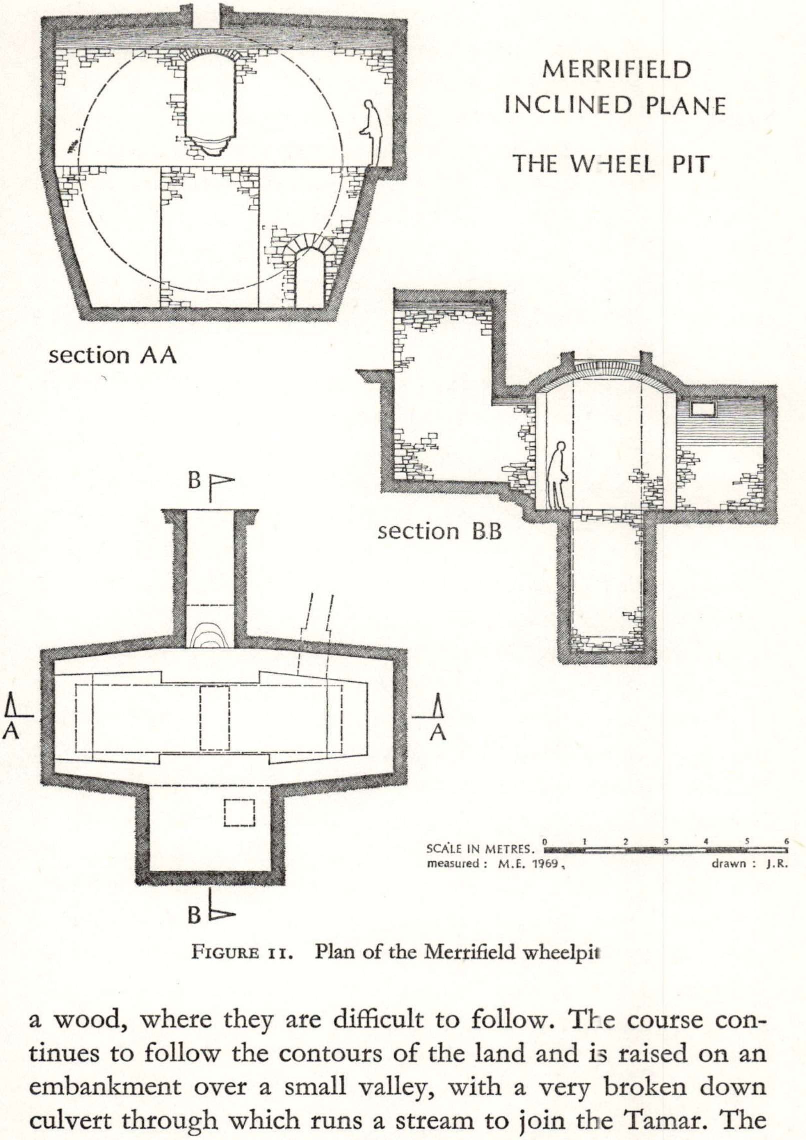

FIGURE 11. Plan of the Merrifield wheelpit

a wood, where they are difficult to follow. The course continues to follow the contours of the land and is raised on an embankment over a small valley, with a very broken down culvert through which runs a stream to join the Tamar. The towpath continues to be well defined and in the copse short of the Crowford road is the Bude 11 milepost.

The canal route takes a sharp turn south-east of East Balsdon Farm and crosses over the streams which join east of the culvert and flow into the Tamar. There is water in this portion of the canal but the course is difficult to follow; there are the footings of an occupation bridge and the canal takes a meandering course over another stream with a culvert in fine condition. North-east of Trepoyle Farm are the remains of 'Iron Bridge', an occupation bridge with very well preserved footings but from which the iron was sold off in the 1930s. Along the next stretch, which displays at intervals the Bude 13, 13¼ and 13½ mileposts, there are the remains of two lay-bys and of a substantial stop-lock for repairs, and the towpath continues in good condition to Tamerton wharf.

The canal crossed the road at a point between Tamerton bridge and the village by an aqueduct, of which only the abutments remain; it was a rather low aqueduct and at times there was trouble when high waggons were unable to pass beneath it and the road was excavated for extra height, necessitating constant repairs. The wharf buildings are now converted to bungalows, the one on the north side of the road having been used for coal and that on the south side for sand.

Also on the south side are the remains of a considerable lay-by, after which the towpath becomes impassable and the canal ploughed in. There is a bend in the line near Tamerton Town, the remains of a footbridge and the Bude 15 and 15¼ mileposts, and presently, continuing along a hillside route overlooking the Tamar valley, the Tamerton incline is reached.

Here the entrance to the wheelpit is still in existence, in size and position exactly similar to that at Merrifield, but the wheelpit itself has collapsed. There are the remains of two channels on either side of a central 'island', probably the channels in which the tub-boats waited until they were sent down the incline or drawn out over the sill, which is still present, into the upper level. Of similar dimensions to Merrifield the plane itself is very overgrown and a former culvert at the bottom is completely destroyed.

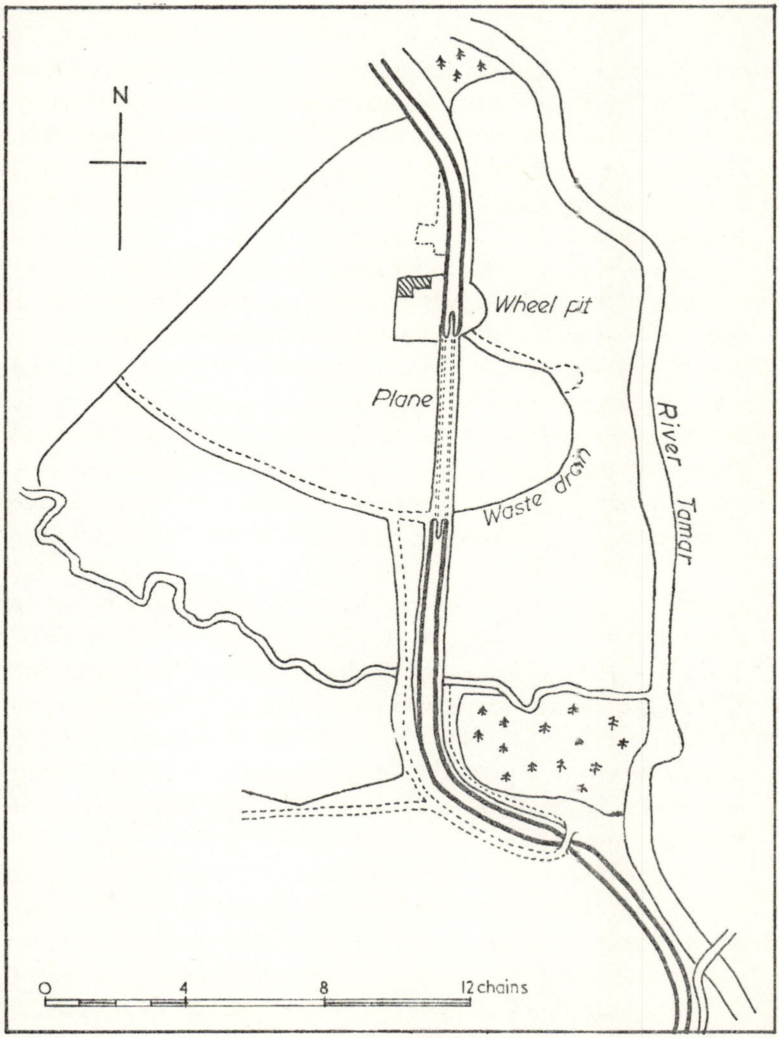

FIGURE 12. Tamerton inclined plane 1842

Continuing southwards, at times coming very close to the Tamar, the canal route runs mostly through woods, mainly following the contours except where crossing tributary streams, and for about a mile in Bradridge Copse is impenetrable. Often just the bed of the canal is discernible but north of Boytonbridge the towpath reappears and, though there is some obliteration of the canal by a Forestry Commission road through Braggshill Wood, is more or less continuous to the Boytonbridge road.

The road crosses the canal route at Boytonbridge without any sign of the bridge which formerly carried it. The area of the canal basin and the wharf here is still clearly apparent but no canal buildings remain; there is, however, a bound stone, and also a chained anchor stone, incorporated in the boundary wall of a modern house. Along the next stage both towpath and canal bed are clearly defined and the 18½ milepost remains; there are two very well preserved culverts, one before and one after a bridge which, like the one at Burmsdon Farm, straddles both canal and towpath.

After the Bude 19 milepost the route follows the contours through Colehill Wood to the destroyed aqueduct over the Tamartown road. It is then well defined to the top of the Werrington or Bridgetown inclined plane, which retains its wheelpit, identical in dimensions to that at Merrifield, and is unusual in that it crosses the public highway by a stone-arched bridge. The wharfinger's house still stands, the surroundings of the plane now constituting a farm yard.

The Bude 20 and 20¼ mileposts are still in situ on the next section, which runs through fields, and also the 20½ post just past an apparent occupation bridge to Bridgetown Farm. Crossing the Tala Water just before it enters the Tamar is an aqueduct with towpath and well-preserved parapet on its south side, a replacement for the original aqueduct which was carried away by a flood in 1835. Continuing overgrown, the route bears again towards the road; there are no signs of the bridge which formerly carried the road over the canal at this point, but in a

Page 121 (*above*) An old tub-boat wheel, held in a piece of metal rail salvaged from the Vealand inclined plane; (*below*) Scene at the foot of the Marhamchurch inclined plane at the turn of the century with the canal choked with weeds and rotting boats. The plane can be seen in the background, sloping up to the sky-line, with the iron foundry at the bottom

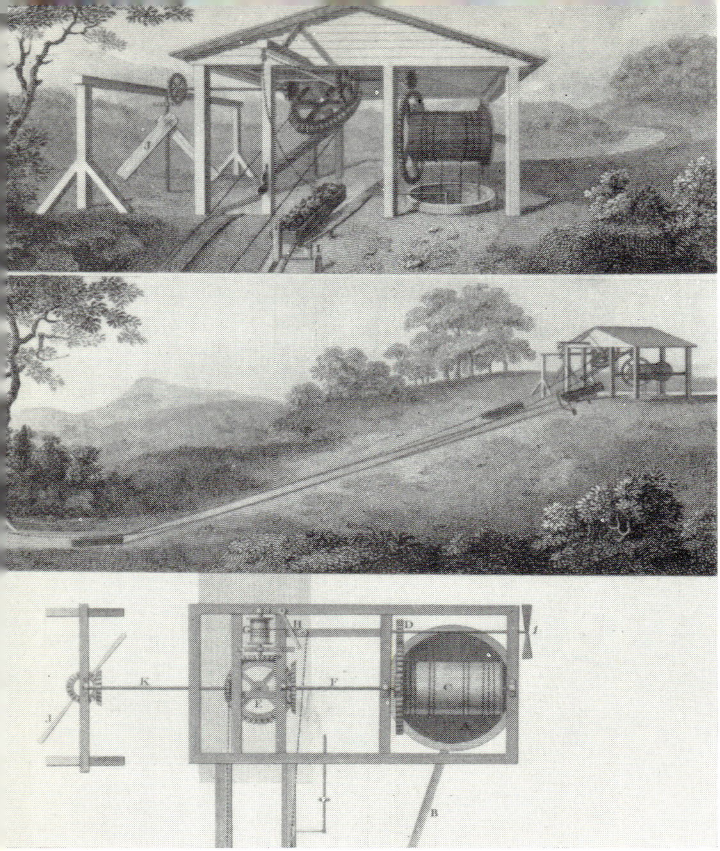

Page 122 (*left*) An illustration from Fulton's *Treatise on Canal Navigation* written in 1796. The design for the Hobba cott inclined plane w based on this principl (*below*) The Hobbacot inclined plane as it now appears. The for- mer keeper's cottage on the right at the to the larger building is close to the pits dowr which the buckets descended to provide power for drawing tl loaded tub-boats up the plane. Behind the buildings is the site o the lay-by where the tub-boats waited

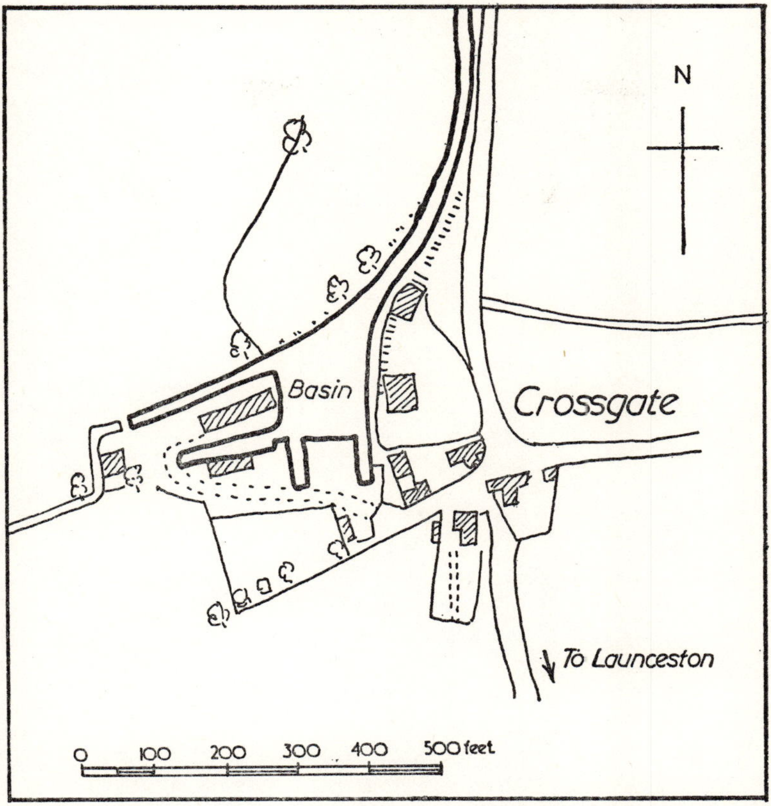

FIGURE 13. Druxton wharf 1884

copse on the other side the canal route can again be discerned. After a short distance it emerges at the terminal basin of Druxton wharf at Crossgate; here the site of the large basin, though now grassed over, is well defined, and there are several canal buildings still in existence—the former wharfinger's house, stables, and a warehouse of typical construction.

Contemplating the remains of the Bude Canal one tries to picture the contrasting scene of a century ago, the cared-for waterways, the busy wharves and terminals and the engineering realities of the inclined planes, noisy with the clanking of metal

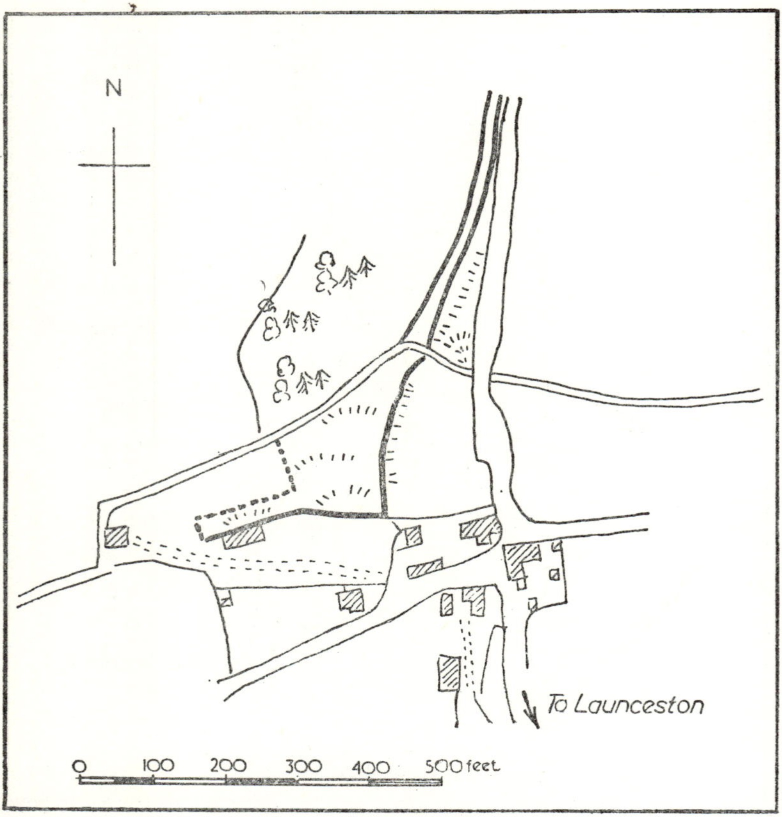

FIGURE 14. Druxton wharf 1953

and the rush of water. How, in some remote reach, a strong ripple of water, which, it is said, preceded oncoming craft like a miniature Severn bore, would herald the approach of a train of tub-boats, presently to come into view around a bend. The boats would have to wait, perhaps, to take turn for descent or ascent over the next plane, while boatmen and plane keeper exchanged topics of local gossip or discussed the state of the trade, and then, the plane negotiated, would go on again along the next sylvan stretch, eventually to disembark their tonnages of sand, coal or whatever at the appropriate wharf, where maybe a farmer's waggon would be drawn up to receive a load. Those who operated the canal were mostly of the same stock as the people who live and work in the district today and must have differed from them very little. The men and women of a generation or two ago could have told us much from their own experiences, but eighty years is a long time to remember, and at best it is only at second-hand that we can hear such few eye-witness accounts as survive.

And so the Bude Canal takes its place as one of the minor wonders of the past. Slowly, as the years slip by, the vestiges of its inland route sink further into obscurity, under the influences of vegetative growth, weathering, agriculture and replanning, in a continuing process which has already proceeded over the greater part of a century. Will the time come when even that which now remains is lost for ever, in significance if not in substance? Few newcomers to the district would immediately recognise—or expect—any of the still-remaining remotely-situated inland features to be associated with a canal, and many present-day dwellers in these parts of Cornwall and Devon are unaware that such a system ever existed. The forces of nature are insidious; those of man can be ruthless, often of necessity, but sometimes through ignorance. Where facts are written down more and more people can become aware of them and perhaps, in this way, even though signs in the countryside may vanish, something of the canal's past and of its atmosphere over those inland stretches may live on.

Bibliography

Bere, Rennie. 'The Nineteenth Century and the Bude Canal', *Post and Weekly News*, 7 July 1962.

Dyer, J. C. MS *History of Bude Harbour and Canal Company* (1901).

Finberg, F. P. R. *Tavistock Abbey* (Cambridge, 1951), 89.

Fulton, Robert. *Treatise on Canal Navigation* (1796), 39–52 and 76–81.

Hadfield, Charles. *The Canals of South West England* (Newton Abbot, 1967).

Leach, Edmund. *A Treatise of Universal Inland Navigation* (1791), 71–5 and 135–45, British Museum.

Marshall, W. *The Rural Economy of the West of England* (2 vols, 1796).

Phillips, John. *History of Inland Navigation* (1792), 267–9, British Museum.

Priestley, Joseph. *Historical Account of Navigable Rivers, Canals and Railways throughout Great Britain* (1831), 110–15, British Museum.

Vancouver, C. *General View of the Agriculture of Devon* (1808).

Notes

NOTES TO CHAPTER 1 (*pages 11–31*)

1. Journal of the House of Lords, 16 May 1774.
2. British Museum, Smeaton's *Reports*, 1812.
3. Report to the Subscribers to the proposed Tamar Manure Canal 1795, Devon Record Office.
4. Report of John and George Nuttall to Meeting at Holsworthy 25 October 1793, Cornwall County Record Office, Truro.
5. Chapman, William. *Observations on Various Systems of Canal Navigation*, 1797.
6. Gilbert, C. S. *Historical Survey of the County of Cornwall* (1820), 2, 539. Truro Museum.
7. Report to Subscribers by James Green, 14 April 1818.
8. Report of the Committee appointed to conduct a survey, 15 April 1818.
9. *Exeter Flying Post*, 22 October 1818.
10. Russell Estate Papers, Devon Record Office.

NOTES TO CHAPTER 2 (*pages 32–47*)

1. Hadfield, Charles. 'James Green as Canal Engineer', *Journal of Transport History* 1 no 1 (May 1953).
2. *Exeter Flying Post*, 5 August 1819.
3. Devon Record Office.
4. *Exeter Flying Post*, 17 July 1823.
5. *Exeter Flying Post*, 31 July 1823.
6. Minutes of the Exchequer Bill Loan Commissioners, Public Record Office.
7. *Exeter Flying Post*, 12 May 1824.

NOTES TO CHAPTER 3 (*pages 48–75*)

1. Bude Canal Book of Reference of Waste Lands, 1840–41.
2. Gilbert, Davies. *Cornwall Parochial History* (1838).
3. *West Briton*, 15 January 1841.
4. Report to the Exchequer Bill Loan Commissioners by J. Walker, 31 May 1838.

5 & 6. Minutes and Reports of the Bude Harbour & Canal Company.

NOTES TO CHAPTER 4 (*pages* 77–100)

1. Robbins, A. F. *Launceston Past and Present*, 1888.
2. £2 per share of the 1,811 (each of nominal value £50) distributed between 130 shareholders. In October 1898 it was reported that Edward Mucklow owned 525 canal shares, his two sons 10 each.

   Annual reports of the committee of management, reports of visiting committees, minute books, letter book and cash book of the Bude Harbour and Canal Company.

   Diary of James Sleeman.

Authors' Notes and Acknowledgements

THE writing of this book has involved the help of very many people, both private individuals and public bodies.

We are very much indebted to the County Record Offices at Truro and Exeter, and to the House of Lords Record Office for facilities for studying original material. We would like to thank the Institution of Civil Engineers for allowing access to Telford's own notes on his visit to the Bude Canal; and the Public Record Office for arranging for the Minute Books of the Exchequer Bill Loan Commissioners to be available.

Particularly we thank Mr F. C. Dredge, Clerk of the Council, for allowing us to spend so many hours studying the material in his care at the Council Offices at Bude.

Our thanks are due also to Mr Trebe Barrett, Mr Basil Bate, Mr R. M. Bere, Mr B. Dudley Stamp, Mr F. C. Dymond, Mr D. J. Gregory, Mr Bruce Sampson, Mr J. E. Stacey, Mrs Sweet and Mrs Wilfrid Wakeley for allowing us to make use of documents in their possession; to Mr Andrew Jewell, Keeper of the Museum of English Rural Life, University of Reading, for his great interest and help in lending so many personal papers; and to the Reverend A. J. Sleeman for lending his grandfather's diary and an original map of the canal from which the plan of Hobbacott incline on page 54 was obtained.

Mr Spencer Howlett and Mrs Vera M. Jennings have put us in touch with a great many people who have proved most helpful in our research. Of these Mr Claude Daniel deserves special thanks for the assistance he gave us in so many ways.

We are very grateful to Mr David St John Thomas and to Mr Charles Hadfield for generously lending us their own detailed files on the canal. The staff of the Map Room at the British Museum have our thanks for supplying copies of the old 6in maps without which the field work on the canal would not have been so easy nor so exciting; and we would like to thank Mr H. L. Douch, Curator of the County Museum, Truro, for his help, and Mr R. L. Hutchings, Curator of the Waterways Museum, Stoke Bruerne, for his kindness in keeping us supplied with relevant material which he found from time to time.

Our grateful thanks are also due to those who have provided us with material for illustrations, whose names are given in the preliminary pages.

In the course of our exploration of the Bude Canal as it is today we have met with nothing but kindness, interest and hospitality. It will be impossible to mention everyone by name, nevertheless Mr Barriball, Mr Cobbledick, Mr S. P. Grylls, Mr Richard Heard, Mr and Miss Moore and Mrs Petherick deserve our special thanks.

Finally, we express our indebtedness to our respective husbands, with whom our interests in the canal and the fascination of its investigation have been completely shared.

APPENDIX 1

Viewing Points on Public Roads

Ordnance Survey 1in to mile, Sheet 174. Bude

Sea Lock to Helebridge (pages 104–6)

1. 207062 Falcon bridge. Lower and upper basins with contemporary canal buildings. Level crossing obliterated by new County bridge.
2. 211048 Roddsbridge. First lock can be seen to the south.
3. 215037 Helebridge. Overgrown wharf, Marhamchurch incline, engineer's house, old warehouse.

Helebridge to Venn (pages 106–9)

4. 224037 Marhamchurch–Stratton road. Line of canal visible.
5. 231046 Near Cann Orchard Farm. Clear view of canal.
6. 233047 Minor road near Cann Orchard. Canal overgrown.
7. 264050 Launceston–Kilkhampton road (B3254). Line of canal visible. Site of branch line to Launceston.
8. 276054 Bude–Holsworthy road (A3072). Inspection necessary to find canal.

Venn to Holsworthy (pages 110–13)

9. 296070 Minor road near Lana. Canal overgrown.
10. 313050 Road from Burnard's House. Canal in deep cutting.
11. 332059 Chilsworthy–Holsworthy road. New bridge, canal bed visible.
12. 343053 Manworthy cross road. View of canal across fields to Stanbury wharf.
13. 348051 Stanbury wharf. Wharfinger's cottage, warehouse.
14. 360057 Blagdonmoor wharf. Contemporary farm buildings, stables, warehouses and remains of tunnel excavations.

Feeder Arm and reservoir (pages 113–14)

15. 294073 Brendon bridge. New bridge, good view of canal.
16. 288082 Puckland bridge. New bridge, good view of canal.
17. 288083 Moreton bridge. Bridge with original cast iron arches still in position.
18. 295087 Dexbeer bridge. New bridge, good view of canal.
19. 298103 Virworthy wharf, Soldon Cross–Kilkhampton road. Wharfinger's cottage, overgrown lay-by, small storehouse.
20. 291114 Tamar Lake, Alfardisworthy–Kilkhampton road. Old bridge, man-made embankment between reservoir and river.

The Launceston branch (pages 114–23)

21. 271028 Moorside–Bridgerule road. Footings of old aqueduct, anchor stone.
22. 267026 Borough Cross–Bridgerule road. No sign of canal.
23. 269024 Newacott–Bridgerule road. No sign of canal.
24. 272014 Whitestone & Bridgerule station. Good view of Merrifield incline.
25. 286992 East Balsdon–Crowford bridge road. Canal overgrown. Good view from bridge over the River Tamar of the line of canal.
26. 316973 North Tamerton–Holsworthy road. Footings of aqueduct, old buildings, site of wharf.
27. 326922 Boyton–Chapman's Well road. Well-marked towpath and site of old wharf.
28. 334908 Minor road at Tamartown. Footings of aqueduct.
29. 338905 Werrington incline. Incline over road and keeper's house.
30. 342892 Bridge over Tala Water. View of aqueduct over Tala Water.
31. 342887 Bridgetown–Crossgate road. Bed of canal just visible.
32. 342884 Druxton wharf. Wharfinger's cottage, warehouse, stables, overgrown basin.

The Inclines

1. 218037 Marhamchurch.
2. 244048 Hobbacott Down.
3. 287066 Vealand.
4. 272018 Merrifield.
5. 325957 Tamerton.
6. 338905 Werrington.

APPENDIX 2

Summary of Facts about the Bude Canal

Statutory Name

Bude Harbour and Canal Company.

List of Acts relating to the Canal

1774. 14 Geo III c 53. Bude Canal Act.
1819. 59 Geo III c 55. Bude Harbour and Canal Act.
1865. 28–29 Vict c 149. The Okehampton Railway (Extensions to Bude and Torrington) Act 1865.
1865. 28–29 Vict c 263. The Bude Canal and Launceston Junction Railway Act 1865.
1891. 54–55 Vict c 75. Bude Harbour and Canal (Further Powers) Act.
1901. 1 Ed VII c 258. Bude Harbour and Canal Abandonment Act.
1960. 8–9 Eliz II c 28. Bude Stratton UDC Act.

Length of original Canal

35½ miles.

Authorised terminal points

Bude—Thornbury—Druxton bridge.

Course of Canal as built

Wide barge canal from Bude to Helebridge.
Narrow tub-boat canal from Helebridge to Blagdonmoor wharf near Holsworthy, with a feeder from the Tamar Lake and a branch to Druxton near Launceston.

Date of commencement of construction—23 July 1819.
Date of completion of Holsworthy line, feeder from reservoir, and Launceston line to Tamerton bridge—8 July 1823.
Date of commencement of construction of Launceston line to Druxton—May 1824.
Date of completion of Launceston line to Druxton—1825.
Date of closure of tub-boat canal—14 November 1891.
Date of final abandonment of works—1901.

Authorised Capital
£95,000.

Number of Locks—Three.

Bude sea lock	Length 114ft Width 33ft Depth of water on cill 9ft 6in
Roddsbridge lock	Length 63ft Width 14ft 7in Depth of water on cill 5ft 6in
Whalesborough lock	Length 63ft Width 14ft 7in Depth of water on cill 5ft 6in

Number of Inclines—Six.

Marhamchurch	120ft rise	836ft long
Hobbacott Down (Thurlibeer)	225ft rise	935ft long
Vealand	58ft rise	500ft long
Merrifield	60ft fall	360ft long
Tamerton	59ft fall	360ft long
Werrington (Bridgetown)	51ft fall	259ft long

Size of boats used on the Canal

Barges	Length	50ft
	Width	13ft
	Draught	3ft 6in
	Capacity	20 tons
Tub-boats	Length	20ft
	Width	5ft 6in
	Draught	1ft 8in
	Capacity	4 tons

APPENDIX 3

Bude Harbour and Canal, Arrivals and Sailings of Vessels

THE following facts are given in three volumes which cover the movements of vessels from July 1824 to December 1900.

Date of arrival. Vessel's name and Master's name, how rigged, number of tons, where belonging, lading inwards, date of entering basin, date of leaving basin, lading outwards and time of sailing.

No information is given of the tonnage of the cargo nor of its port of origin.

The first volume which covers the period July 1824 to May 1835 gives the following information.

Where belonging Appledore, Barnstaple, Berkley, Bideford, Boscastle, Bridgewater, Bristol, Bude, Cardigan, Cardiff, Chepstow, Clovelly, Dartmouth, Dundee, Exeter, Falmouth, Fowey, Gloucester, Gweek, Hartland, Ilfracombe, Kidwelly, Laugharne, London, Looe, Lydney, Minehead, Newport, Padstow, Penzance, Plymouth, Port Isaac, Portquin, Portreath, Portsmouth, Salcombe, St Ives, Swansea, Waterford, Weston super Mare, Wexford, Whitby.

Lading inwards Coal, Bristol Stone, Slate, Timber, Merchandise, Limestone, Shafts for Company, Culm, Iron, Porter, Salt, Tobacco, Earthenware, Sand for casting iron, Furniture, Bricks, Grindstones, Cider, Tallow, Herrings, Castings, Fish, Deal, Granite.

Lading outwards Ballast, Grain, Bark, Timber, Potatoes, Malt, Corn, Manganese, Bath stone, Household goods, Iron goods.

In the second volume covering the rest of 1835 to the end of 1859 the entries for 'where belonging' are less varied, and only the following ports are mentioned.

Barnstaple, Beaumaris, Bristol, Cardiff, Clovelly, Dartmouth, Exeter, Gloucester, Hartland, Ilfracombe, Llanelly, Looe, Newport, Padstow, Penryn, Plymouth, Swansea.

By the end of 1838 most vessels came from Padstow. In 1852 the port of registration was mainly Bideford.

During this period the inward cargoes are similar to those mentioned in the first volume with the addition of guano from 1847.

Lading outwards Ballast, Grain, Bark, Timber, Poles, Iron, Malt, Hay, Tiles, Copper ore.

In the third volume it appears that all vessels 'belonged' to Bude.

Lading inwards In addition to the cargoes mentioned in the first two volumes, are the following—Bone ash, Guano, Superphosphates, Oats, Cement, Gravel, Rails, Chairs, Sleepers, Crackstone, Pipes and Phosphates.

Lading outwards Nearly all vessels left in ballast with occasional cargoes of grain and bark. In 1861 iron ore was exported and in 1872 there were several cargoes of scrap iron.

APPENDIX 4

Trade Figures from 1825 to 1900

Number of Cargoes (Not tonnage)

Year	*Coal*	*Lime-stone*	*Goods*	*Culm*	*Slate*	*Salt*	*Timber*	*Iron*	*Bricks*	*Guano*	*Super-phos-phate*	*Phos-phate*	*Sleepers, Rails etc*
1825	77	4	16	3	27	4	5	–	2	–	–	–	–
1826	61	4	16	6	12	2	4	1	1	–	–	–	–
1827	71	11	17	21	21	4	6	5	2	–	–	–	–
1828	81	20	13	16	14	3	7	4	3	–	–	–	–
1829	81	34	14	24	20	4	4	2	4	–	–	–	–
1830	73	45	14	22	13	3	9	2	3	–	–	–	–
1831	85	48	16	35	27	6	9	4	3	–	–	–	–
1832	132	61	16	28	15	6	9	4	4	–	–	–	–
1833		Missing											
1834		Missing											
1835	134	37	8	19	10	3	14	1	3	–	–	–	–
1836	107	31	10	27	11	5	6	1	1	–	–	–	–
1837	149	46	10	43	10	6	15	1	2	–	–	–	–
1838	127	38	12	44	8	4	9	1	2	–	–	–	–
1839	174	47	12	46	9	8	15	2	1	–	–	–	–
1840	174	34	12	38	6	5	3	3	2	–	–	–	–
1841	208	27	8	41	6	7	6	5	–	–	–	–	–
1842	201	30	9	42	4	5	6	7	2	–	–	–	–
1843	180	19	12	27	5	6	6	2	–	–	–	–	–
1844	205	30	9	29	8	9	6	6	2	–	–	–	–
1845	189	39	7	31	5	4	6	2	2	–	–	–	–
1846	225	23	10	31	6	7	4	1	1	–	–	–	–
1847	165	32	11	42	5	3	7	2	3	1	–	–	–
1848	205	24	13	34	6	6	10	4	3	6	–	–	–
1849	166	14	10	30	4	8	6	3	1	1	–	–	–
1850	138	10	9	32	5	6	3	3	2	1	–	–	–
1851	152	19	11	29	4	8	2	3	–	5	–	–	–
1852	149	14	10	33	5	3	4	3	–	7	–	–	–
1853	155	12	17	31	7	3	9	–	1	11	–	–	–
1854	115	20	14	40	5	6	7	–	2	12	–	–	–
1855	147	22	17	42	6	8	10	–	1	13	2	–	–
1856	124	13	15	42	5	10	7	–	1	21	–	–	–
1857	127	13	20	29	8	14	4	–	2	13	6	–	–
1858	132	21	20	40	7	9	16	2	1	21	4	–	–

Year	*Coal*	*Lime-stone*	*Goods*	*Culm*	*Slate*	*Salt*	*Timber*	*Iron*	*Bricks*	*Guano*	*Super-phos-phate*	*Phos-phate*	*Sleepers, Rails etc*
1859	152	19	30	27	9	7	7	2	1	17	6	–	–
1860	164	21	26	36	9	8	10	2	3	24	2	–	–
1861	166	24	20	24	10	6	9	1	2	13	9	–	–
1862	171	23	16	40	4	5	4	–	2	11	5	–	–
1863	152	18	21	28	4	4	5	–	1	8	7	–	–
1864	161	24	19	36	6	6	3	–	2	5	1	5	–
1865	136	20	15	31	4	6	9	–	3	7	2	8	–
1866	128	23	13	12	2	3	1	–	8	6	–	5	–
1867	153	28	10	21	7	4	3	–	1	5	1	5	–
1868	118	17	9	23	6	5	3	–	3	10	–	7	–
1869	126	36	15	17	3	3	4	–	2	7	–	8	–
1870	117	33	16	13	5	4	7	–	4	7	–	8	–
1871	122	37	19	19	4	4	2	–	2	5	4	13	–
1872	115	25	16	22	3	4	6	–	2	3	–	16	–
1873	106	35	16	12	4	2	5	–	4	2	–	25	–
1874	109	23	16	13	4	5	4	–	3	1	–	25	–
1875	109	21	11	8	6	3	5	–	2	2	–	22	3
1876	99	20	18	3	6	3	5	–	4	–	–	27	34
1877	117	27	17	10	2	4	8	–	7	1	–	27	6
1878	122	36	23	5	10	2	6	–	4	1	–	26	3
1879	96	22	17	5	4	4	4	–	2	–	–	12	2
1880	88	9	12	4	3	4	4	–	2	–	–	13	–
1881	92	21	8	4	4	2	6	2	4	–	–	16	–
1882	94	28	9	5	4	3	7	1	4	–	–	15	–
1883	108	16	13	4	3	2	7	1	2	–	–	20	–
1884	111	17	14	5	4	3	8	1	4	–	–	18	–
1885	116	21	13	4	6	3	10	–	3	–	–	23	–
1886	116	13	11	5	3	4	8	–	4	–	–	27	–
1887	103	17	17	4	5	4	10	1	2	–	–	33	–
1888	116	15	27	3	5	3	9	1	8	–	–	33	–
1889	113	18	24	1	10	5	10	2	5	–	–	32	–
1890	96	19	18	5	4	2	15	–	4	–	–	36	–
1891	87	14	18	4	4	5	12	1	7	–	–	26	3
1892	103	17	17	5	5	2	11	3	6	–	–	26	–
1893	87	16	21	3	6	1	7	1	5	–	–	29	–
1894	104	15	23	3	3	4	8	1	13	–	–	32	–
1895	99	10	23	3	3	3	7	2	6	–	–	35	–
1896	91	18	24	4	3	4	12	2	11	–	–	34	1
1897	99	14	23	2	5	5	13	4	10	–	–	32	19
1898	89	9	19	4	2	1	11	1	15	–	–	27	4
1899	84	10	15	3	3	2	8	1	13	–	–	29	–
1900	67	12	15	3	2	1	3	–	12	–	–	29	–

Page 139 (*above*) Old buildings and the site of the former basin at the top of the Hobbacott inclined plane; (*right*) Present-day upward view from the foot of the former Hobbacott inclined plane. The biggest and most troublesome of the canal's six inclines, it was 935ft in length and raised the tub-boats to a height of 225ft

Page 140 (*above*) The aqueduct which carried the canal over the River Tamar at Burmsdon; (*below*) A pair of anchor stones with chains attached, seen at Stanbury wharf

APPENDIX 5

Number of Vessels Trading at Bude, 1825-1900

Year	*Jan*	*Feb*	*Mar*	*Apr*	*May*	*June*	*July*	*Aug*	*Sept*	*Oct*	*Nov*	*Dec*	*Total*
1825	6	10	10	16	24	16	27	15	14	6	–	7	151
1826	6	1	9	13	14	20	19	19	13	3	6	8	121
1827	6	10	7	21	17	26	22	24	13	7	8	1	162
1828	2	13	8	11	25	25	23	14	10	19	9	3	162
1829	10	4	21	12	18	26	16	15	20	19	18	7	186
1830	2	3	15	17	25	26	21	26	11	22	7	7	182
1831	13	10	11	16	47	21	43	26	22	7	8	14	238
1832	12	10	21	23	28	18	48	30	39	18	16	6	269
1833	12	6	15	13	33	25	39	43	15	12	Missing		
1834		Missing			36	27	34	26	37	24	5	6	
1835	4	2	10	28	27	38	40	25	16	17	21	7	235
1836	7	11	3	14	33	30	19	32	17	22	8	5	201
1837	7	13	29	16	38	41	34	40	21	13	17	12	281
1838	7	7	21	23	34	32	27	28	32	15	19	7	252
1839	13	13	20	27	31	41	43	38	28	39	12	7	312
1840	1	9	30	34	39	29	27	27	24	33	9	9	271
1841	12	9	21	31	34	39	41	28	39	21	21	10	306
1842	10	13	8	54	29	39	34	39	30	37	8	3	304
1843	3	18	17	32	36	33	23	35	38	13	3	7	258
1844	12	8	22	43	51	31	35	16	34	18	16	13	299
1845	3	17	21	35	43	22	36	41	18	23	21	8	288
1846	9	19	28	39	36	33	34	36	25	22	15	19	315
1847	15	11	17	20	27	34	44	36	25	16	15	10	270
1848	22	11	14	25	45	36	36	27	50	19	17	7	309
1849	8	13	21	15	18	48	34	27	37	19	4	7	251
1850	4	4	21	8	16	29	24	28	26	26	14	8	208
1851	–	22	16	21	25	19	26	23	23	25	20	11	231
1852	–	14	30	27	28	18	31	18	26	28	9	3	231
1853	6	18	15	25	32	27	22	29	28	17	16	21	256
1854	4	14	19	27	25	26	30	20	19	14	20	3	221
1855	24	13	12	21	30	29	31	22	28	25	23	13	271
1856	11	8	18	21	34	22	20	16	30	30	16	16	242
1857	8	13	22	24	31	21	17	33	17	16	32	8	242
1858	4	26	23	35	27	29	30	29	26	22	25	–	276

Year	*Jan*	*Feb*	*Mar*	*Apr*	*May*	*June*	*July*	*Aug*	*Sept*	*Oct*	*Nov*	*Dec*	*Total*
1859	15	15	25	29	38	37	23	17	15	30	18	12	274
1860	8	27	15	31	37	24	44	26	31	14	24	20	301
1861	15	23	19	42	35	29	19	20	26	31	11	18	288
1862	9	28	26	33	28	29	30	33	29	17	19	7	288
1863	11	10	20	21	31	32	30	30	18	15	10	23	251
1864	13	15	34	27	42	21	27	12	18	32	13	15	269
1865	15	6	18	40	29	26	26	20	21	13	12	15	241
1866	10	7	23	21	24	23	21	13	11	24	12	13	202
1867	11	10	22	14	36	26	32	21	18	18	20	11	239
1868	9	8	20	23	22	21	25	10	23	17	16	8	202
1869	11	5	22	31	31	23	21	16	7	24	9	21	221
1870	19	7	30	20	24	17	19	15	14	9	9	29	212
1871	18	15	21	23	38	23	15	17	25	11	22	8	236
1872	13	17	12	27	30	15	26	19	13	17	14	8	211
1873	9	8	32	25	20	20	20	14	16	22	12	11	209
1874	9	14	17	24	18	24	19	13	21	26	11	10	206
1875	–	23	18	21	22	16	24	18	9	14	15	14	194
1876	11	12	20	22	27	25	23	14	13	17	22	13	219
1877	6	3	32	26	27	25	17	30	27	13	10	15	231
1878	13	10	29	25	32	29	22	22	12	10	11	22	237
1879	5	12	17	16	17	23	12	9	12	19	10	14	166
1880	9	7	11	18	18	13	12	8	7	12	14	6	135
1881	11	11	17	12	18	11	12	9	23	13	8	8	153
1882	9	16	12	18	21	19	13	11	17	19	4	8	167
1883	10	8	17	19	23	20	9	13	15	8	6	20	168
1884	8	16	10	24	19	21	13	14	11	11	19	14	180
1885	11	5	22	15	28	19	20	10	11	12	21	13	187
1886	15	13	13	26	19	15	11	13	15	15	19	12	186
1887	13	14	19	22	23	15	8	6	11	21	15	18	185
1888	8	24	19	21	28	21	11	12	17	17	9	22	209
1889	8	14	27	19	20	27	16	11	18	15	25	6	206
1890	11	15	14	25	32	7	17	11	19	10	5	14	180
1891	10	19	12	22	17	14	7	13	13	15	16	7	165
1892	8	17	24	22	12	18	16	9	13	23	11	11	184
1893	12	8	27	21	12	11	14	16	9	16	16	6	168
1894	5	10	19	19	26	17	14	17	21	17	13	11	189
1895	10	14	12	18	24	17	13	13	22	22	9	10	184
1896	10	19	9	25	24	18	5	14	8	21	21	14	188
1897	20	11	8	27	26	18	14	22	17	29	16	17	225
1898	17	11	24	17	20	16	15	13	19	7	13	7	179
1899	11	5	20	15	25	18	18	10	7	16	13	8	166
1900	7	13	17	14	19	18	16	10	10	12	6	3	145

APPENDIX 6

Vessels Trading with Bude at 20 year intervals, 1830-90

1830

Vessel's Name	*Master's Name*	*How rigged*	*No of tons*	*Where belonging*	*No of sailings*
Agnes	Lewis	Sloop	38	Swansea	1
Betsy	Knight	,,	32	Padstow	1
Bude Packet	Drew	,,	44	,,	14
Catherine	Symons	,,	35	Appledore	1
Ceres	Knowles	,,	57	Dartmouth	8
Eleanor	Tucker	,,	35	Padstow	22
Eliza	Worth	Boat	5	Port Isaac	4
Ferret	Strout	Sloop	23	Padstow	2
Friends	James	,,	52	Penzance	1
Friends	Whitfield	,,	29	Bude	19
Friendship	Bickham	,,	60	Bideford	4
Friendship	Holland	,,	45	Swansea	2
Harmony	Chilcut	,,	10	Padstow	1
Henrietta	Moxley	,,	54	Newport	2
Hope	Adams	Boat	4	Port Isaac	1
Hope	Buckler	Sloop	36	Clovelly	5
Hope	Reese	,,	35	Cardigan	1
Hudscott	Rowe	,,	77	Bideford	2
Jane	Frances	,,	59	St Ives	1
John	Reese	,,	53	Hartland	3
John and Mary	Langford	,,	21	Swansea	1
Kingfisher	Brown	,,	19	Padstow	1
Lanson Castle	Tucker	,,	40	Bude	13
Le Hereaux	Hoskin	,,	15	Padstow	3
Margaretta	Wilson	,,	35	,,	1

Vessel's Name	*Master's Name*	*How rigged*	*No of tons*	*Where belonging*	*No of sailings*
Maria	Metherel	Sloop	62	Newport	9
Mary	Southwood	,,	36	Padstow	13
Ranger	Lee	,,	33	Bideford	1
Rising Sun	Lewis	,,	37	Padstow	11
Roebuck	Finch	,,	51	Hartland	4
Rover	Berry	Brig	92	Bideford	1
Sally	Lee	Sloop	20	,,	3
Sisters	Cook	Schooner	49	Padstow	15
Sprightly	Briton	Sloop	47	Hartland	2
Stormstone	Smith	,,	60	Bideford	1
Success	Clemar	,,	68	Looe	1
Surprise	Braund	,,	21	Bideford	2
Susan	Chalk	,,	13	Padstow	3
Three Brothers	Billing	,,	39	,,	1
Raft of Timber			4½		
1850					
Ada	Hutchings	Sloop	34	Padstow	1
Affo	Beer	,,	43	,,	11
Aunda	Cook	,,	8	,,	1
Battreaux Castle	Short	,,	40	,,	2
Bee	Pengelly, Barrett	,,	51	,,	11
Ceres	Lewis	,,	34	,,	9
Charles	Jones	,,	38	,,	1
Corsair	Hart	,,	46	,,	1
Eclipse	Gumyn	,,	59	,,	11
Eleanor	Tucker	,,	25	,,	5
Eliza	Baynon	,,	19	,,	2
Emma	Drew	,,	49	,,	9
Friends	Davey	,,	36	,,	1
Happy Return	Finch	,,	40	,,	1
Hope	Richards	,,	33	,,	1
Lady Acland	Davey	,,	36	,,	16
Lady Mary	Whitfield	,,	33	,,	13
Margaretta	Fish	,,	17	,,	13
Mirre	Pickard	,,	26	,,	19
Sisters	Davey, Barrett	,,	35	,,	9
Sprightly	Beer, Marshall	,,	37	,,	2

Vessel's Name	*Master's Name*	*How rigged*	*No of tons*	*Where belonging*	*No of sailings*
Star	Drew	Sloop	34	Padstow	19
Stuckley	Gill	,,	27	,,	15
William & Betsy	Budge, Hatherly	Schooner	33	,,	14
Winlay	Gill	Sloop	27	,,	4
Yeo Glory	Hawood	,,	33	,,	17
1870					
Ada	—	Sloop	45	Bude	1
Affo	Goman, Davis	,,	37	,,	7
Ant	Stapleton	,,	35	,,	11
Bee	Hallett	,,	49	,,	22
Brittania	Swain	,,	39	,,	2
Brother	Wallis	,,	40	,,	1
Ceres	Petherick	,,	44	,,	8
Coronation	Cook	,,	44	,,	5
Deerhound	—	,,	59	,,	1
Eling	Hinds	,,	84	,,	6
Elizabeth	Lashbrook	,,	42	,,	3
Elizabeth Martin	Maynard	,,	57	,,	10
Elizabeth Scown	Lashbrook, Elliott	,,	61	,,	9
Emma	Drew	,,	43	,,	6
Endeavour	Goman	,,	45	,,	1
Glendirgle	Bennett	,,	62	,,	1
Hawk	Barrett	,,	44	,,	9
John Hedley	Hill	,,	49	,,	2
Julia	Brent	,,	60	,,	1
Lady Acland	Elliott, Wonnacott	,,	47	,,	15
Lady of the Lake	Peardon	,,	59	,,	13
Livinia	Finison	,,	199	,,	1
Mirre	Bate	,,	22	,,	18
Nancy	Ridgman	,,	47	,,	10
Newport	Bate	,,	89	,,	9
Reward	Anderson	,,	44	,,	1
Sir T. Acland	Whitfield	,,	40	,,	20
Stag	Burnard	,,	34	,,	1

Vessel's Name	*Master's Name*	*How rigged*	*No of tons*	*Where belonging*	*No of sailings*
Stuckley	Barrett	Sloop	27	Bude	13
Vesta	Oats	,,	41	,,	1
Volant	Addison	,,	42	,,	1
1890					
Agnes	Roland	Sloop	58	Bude	9
Alarm	Rosevear	,,	32	,,	1
Alford	Cook	,,	65	,,	6
Annie Davey	Howell	,,	63	,,	3
Ant	Lucas, Hinds	,,	48	,,	7
Boconoe	Penwarden	,,	36	,,	20
Ceres	Petherick	,,	44	,,	10
Charles Francis	Lewis	,,	76	,,	1
Duwin	Baldson	,,	49	,,	1
Daviewyn	Slade	,,	49	,,	1
Echo	Mitchell	,,	36	,,	1
Eling	Hallett	,,	84	,,	1
Elizabeth	Lashbrook	,,	50	,,	9
Endeavour	Sluggatt	,,	38	,,	12
Faith	Rook	,,	51	,,	12
Fanny	Hare	,,	49	,,	1
Friendship	Stephens	,,	59	,,	1
Hawk	Brinton	,,	36	,,	21
Infanta	Batten	,,	56	,,	1
Jessamine	Maynard	,,	48	,,	2
John and Ann	Drake	,,	38	,,	2
Joseph & Thomas	Shazel	,,	47	,,	2
Katie	Kelly	,,	69	,,	1
Lady Acland	Barrett	,,	43	,,	17
Lady of the Lake	Stephens	,,	59	,,	4
Liver	Reeks	,,	44	,,	1
Morfa	Sully	,,	35	,,	1
Purveyor	Stanbury	,,	39	,,	4
Rapid	Webb	,,	48	,,	1
Sarah	Knight	,,	64	,,	1
Sir T. Acland	Hallett	,,	40	,,	7
Stuckley	Brock, Barrett	,,	32	,,	11
Tavy	Mountjoy	,,	69	,,	4

Vessel's Name	*Master's Name*	*How rigged*	*No of tons*	*Where belonging*	*No of sailings*
United Friends	Gard	Sloop	34	Bude	1
Uyre	Evers	,,	57	,,	1
Venerable	Owen	,,	86	,,	1
Wave	Beard	,,	83	,,	1

APPENDIX 7

Summary of Tonnage carried on the Canal between 1837 and 1890

Year	*Culm*	*Coal*	*Sand*	*Goods*	*Total tonnage*
1837	977	2,615	42,059	131	45,782
1838	1,796½	3,575	54,016	233	59,620½
1839	1,509½	4,034	45,692	423	51,658½
1848					52,501
1858					43,211½
1868					53,103½
1878	357½	5,561½	40,732½	1,934¾	48,586¼
1879	4	5,367	40,365	1,022¼	46,758¼
1880	–	3,232½	50,052½	803¼	54,088¼
1881	4	3,510	37,844¾	792	42,150¾
1882	–	3,401½	38,340½	1,586¼	43,328¼
1883	–	3,369	29,642	944	33,955
1884	–	3,640	24,893	1,072	29,605
1885	–	3,370	27,222	1,097	31,689
1886	–	4,097	29,014	1,364	34,475
1887	–	3,813	23,687	1,284	28,784
1888	–	2,718	18,414	1,675	22,807
1889	–	3,044	25,272	2,295	30,611
1890	–	3,148	16,403	2,764	22,315

APPENDIX 8

Summary of Income and Expenditure

	INCOME		EXPENDITURE			
Year	*Harbour, Basin & Wharf Dues*	*Tonnage Dues*	*Salaries*	*Day labour*	*Working Hobbacott*	*Breakwater repairs*
	£ s d	£ s d	£ s d	£ s d	£ s d	£ s d
1823	Not known	1,470 10 5	Not known	Not known	Not known	Not known
1824	,,	2,775 2 8	,,	,,	,,	0 0 0
1825	115 14 0	2,356 1 8	,,	,,	,,	143 0 2
1826	106 1 7	2,655 7 2	,,	,,	,,	21 15 9
1827	131 2 0	3,412 3 2	,,	,,	,,	34 5 10
1828	124 8 6	2,392 13 10	,,	,,	,,	1,037 2 10
1829	114 15 0	2,746 18 8	,,	,,	,,	448 18 8
1830	137 11 11	1,769 18 11	,,	,,	,,	89 0 10
1837	139 7 6	3,005 4 6	289 11 8	636 14 5	271 14 11	0 0 0
1838	143 17 10	4,183 7 4	280 8 4	711 1 5	491 3 4	31 15 6
1839	141 7 10	4,016 14 5	274 3 4	657 17 3	108 18 6	1,314 14 2
1848		3,532 18 0				
1858		3,205 13 9				
1860	249 7 4	Not known	Not known	912 16 7	124 10 6	8 18 6
1861	234 0 6	,,	,,	924 0 11	124 10 6	12 10 0
1862	255 11 8	,,	,,	898 0 0	113 6 6	31 18 0
1863	201 6 2	,,	,,	922 15 6	110 14 6	64 3 8
1864	205 15 10	,,	,,	925 7 7	108 0 0	45 2 4
1865	199 11 5	,,	,,	937 17 6	107 8 0	148 3 8
1866	147 5 10	,,	,,	893 3 8	107 14 6	64 6 0
1867	186 0 2	,,	,,	889 0 1	107 14 6	7 12 0
1868	169 1 4	2,321 19 6	,,	832 0 10	108 1 0	11 2 8
1869	165 4 10	Not known	,,	848 12 5	107 14 6	16 12 0
1870	145 10 10	,,	,,	824 3 1	107 14 6	4 0 0
1871	176 0 2	,,	,,	962 5 8	107 8 0	0 0 0
1872	164 15 7	,,	,,	1,006 6 5	108 1 0	0 0 0
1873	140 14 10	,,	,,	891 13 5	109 0 10	15 16 8
1874	137 3 8	,,	,,	921 0 4	109 18 10	0 0 0
1875	140 17 2	,,	,,	894 0 2	116 8 10	0 0 0
1876	195 10 10	,,	,,	873 17 6	112 18 10	11 8 6

Year	INCOME: *Harbour, Basin & Wharf Dues*	INCOME: *Tonnage Dues*	EXPENDITURE: *Salaries*	EXPENDITURE: *Day labour*	EXPENDITURE: *Working Hobbacott*	EXPENDITURE: *Breakwater repairs*
	£ s d	£ s d	£ s d	£ s d	£ s d	£ s d
1877	193 19 5	Not known	Not known	892 13 4	116 8 10	42 19 0
1878	Not known	,,	,,	Not known	Not known	Not known
1879	191 3 1	1,636 11 1	265 0 0	849 6 9	118 3 2	20 13 7
1880	135 7 5	1,752 18 1	265 0 0	805 11 0	119 3 2	89 0 10
1881	125 8 7	1,414 16 1	267 2 0	871 6 1	118 3 2	34 13 7
1882	Not known	Not known	Not known	Not known	Not known	Not known
1883	133 9 5	1,143 7 8	265 0 0	750 3 7	108 8 10	32 19 8
1884	138 19 2	1,034 19 6	221 10 0	694 17 8	93 14 0	56 8 8
1885	144 18 6	1,086 6 3	207 0 0	655 2 4	93 14 8	95 12 9
1886	166 4 10	1,228 4 8	209 2 0	693 18 7	99 0 8	58 8 6
1887	158 3 0	1,242 15 0	209 2 0	731 2 5	97 14 0	17 3 2
1888	149 6 11	826 9 8	207 0 0	774 4 11	97 13 2	15 13 6
1889	178 16 11	1,093 11 8	207 0 0	725 2 4	97 4 4	23 1 8
1890	172 17 2	827 17 3	207 0 0	771 2 8	93 3 4	13 19 0
1891	154 17 1	527 4 9	207 0 0	755 16 2	98 1 4	86 16 2
1892	Not known	Not known	Not known	Not known	Not known	Not known
1893	174 16 3	17 11 7	159 0 0	136 7 7	0 0 0	45 11 5
1894	164 13 11	16 11 7	97 0 0	166 18 4	0 0 0	Not known
1895	156 9 0	10 10 6	88 16 8	163 19 11	0 0 0	,,
1896	183 17 9	17 18 2	65 0 0	180 12 8	0 0 0	,,
1897	214 0 2	35 16 5	65 0 0	159 17 5	0 0 0	,,
1898	173 0 4	8 18 11	65 0 0	170 4 9	0 0 0	,,
1899	134 6 11	1 0 10	60 0 0	215 3 4	0 0 0	,,
1900	124 10 7	3 14 8	50 0 0	107 17 8	0 0 0	,,
1901	124 3 7	Not known	50 0 0	189 19 5	0 0 0	,,

APPENDIX 9

Profit and Loss Account and Dividends

DURING the last twelve years of the working life of the canal, the profits which were never large, fell sharply. After the closure of the tub-boat canal in 1891 the Company lost money, until in 1897 there was an improvement due to the building of the branch railway line to Bude, when the canal was used for the carriage of sleepers, rails etc to Helebridge.

Dividends paid by the Company on the original £50 shares were negligible. Nothing was paid before 1876.

Year	*Profit*			*Loss*				*Dividend*		
	£	s	d	£	s	d		£	s	d
1876			Not known						10	0
1877			Not known						3	4
1878			Not known						5	0
1879	632	18	10	–	–	–			5	0
1880	485	16	4	–	–	–			6	8
1881	107	8	7	–	–	–			–	–
1882	156	18	0	–	–	–			3	4
1883	31	2	6	–	–	–			–	–
1884	–	–	–	38	13	3			–	–
1885	–	–	–	407	2	7			–	–
1886	217	0	4	–	–	–			–	–
1887	204	9	4	–	–	–			–	–
1888	–	–	–	314	2	9			–	–
1889	213	10	5	–	–	–			–	–
1890	–	–	–	180	7	5			–	–
1891	–	–	–	632	3	6			–	–
1892	–	–	–	556	3	2	(15 months)		–	–
1893	–	–	–	105	2	4			–	–

Year	*Profit*			*Loss*			*Dividend*		
	£	s	d	£	s	d	£	s	d
1894	–	–	–	75	13	6		–	–
1895	–	–	–		12	8		–	–
1896	–	–	–	18	2	7		–	–
1897	100	8	9	–	–	–		–	–
1898	61	11	5	–	–	–		–	–
1899	–	–	–	72	16	7		2	6
1900	14	12	4	–	–	–	4	0	0
							(Proceeds of sale of canal)		
1901	76	0	8	–	–	–		–	–

APPENDIX 10

Company Employees

Engineer

1819–1832	James Green
1832–1876	George Casebourne. £175 pa increased to £200 in 1840
1876–1901	Nicholas Sullivan. £150 pa reduced to £125 in 1883. In 1894 he became consulting engineer at £52 pa

Clerk to Company

1823–	George Thorn
1820–1831	Isaac Whitewood
1824–1832	John Honey, at Bude
1832–	Joseph Cox
1834–1862	James Collom at Bude. £85 pa reduced to £45 pa in 1859
1859–1863	George Lang. £5 16s 8d per month
1863–1899	William Davey at Bude. £7 1s 8d per month
1865–1878	F. Hutchings. £7 10s quarterly plus £2 10s for rent

Clerk to Committee of Management

1820–1837	John Waddington
1838	William Thomas. £30 pa
1859–1864	George William Branscombe. £2 10s per month
1878–1885	J. F. Eldridge
1886–1902	Fred Pollard. £7 10s quarterly plus £2 10s for rent

Canal Employees

1837	Charles Brown, conduiter to Hobbacott Plane, 3s 6d per day with a house
	Thomas Shepeard, assistant to above, 3s per day with a house

1837–1887	John Balsdon, blacksmith to the company
1838	William Dyer, mason, 2s 3d per day
1839	George Bond, wharfinger at Hele, 2s 3d per day
	Thomas Brown Jun, carpenter, 1s per day
	Thomas Smale in charge of Tamerton line, increased from 1s 8d to 2s per day 'as a reward for his attentive conduct'
1840	J. Gourman, lock-keeper
1843	Thomas Smale, plane keeper at Werrington
1850	James Collom, clerk at Bude also has duties as wharfinger at Hele
1856	John Gliddon, wharfinger at Blagdonmoor
1866–1890	John Hutching, wharfinger at Blagdonmoor
1871	J. Smale, plane keeper at Werrington
1877	R. Gerry, care of reservoir and parts of canal
1878	John Andrew, care of reservoir
	James Sleeman, carpenter, 19s per week
1882	Charles Brown Sen retires as plane keeper at Hobbacott
	James Sleeman, plane keeper at Hobbacott
1883	William Found, harbour master
1891	Noah Smale, plane keeper at Werrington
1897	Thomas Found, harbour master

APPENDIX 11

Wrecks involving Vessels Trading at Bude

Date	*Vessel*	*Cargo*	*Where wrecked*
*1862 November	*Mary Elizabeth* of Padstow	Coal from Newport Mon to Bude	Millook, about 7 miles SW of Bude
1864 December	*Mirre* of Bideford	General cargo from Bristol to Bude	Entrance to Bude Haven
1865 February	*Endeavour* of Bideford	Slates from Port Gaverne to Bude	Entrance to Bude Haven
1865 June	*Johnson* of Exeter	Coal from Newport Mon to Bude	Entrance to Bude Haven
1866 March	*Maggie* of Liverpool	Ballast from Bude to Swansea	500yd N of Bude
*1867 June	*Uncle John* of Bideford	Coal from Newport Mon to Bude	Rocks outside Bude breakwater
1868 November	*Ant* of Bideford	Wheat from Bude to Plymouth	Entrance to Bude Haven
1871 March	*Armyn* of Falmouth	Patent manure from Penryn to Bude	Entrance to Bude Haven
1872 November	*Nancy* of Padstow	Coal from— to Bude	Beacon Rock at entrance to Bude Haven
*1873 January	*Ellen Martin* of Bideford	Coal from Newport Mon to Bude	½ mile E of Bude
*1873 March	*Anna Gezina* of Holland	Patent manure from London to Bude	Entrance to Bude Haven
1874 February	*Annie Davey* of Bideford	Manure from Newport Mon to Bude	Bude Harbour
1874 March	*Arrow* of Jersey	Patent manure from Wemyss to Bude	Bude Harbour
1874 March	*Hawk* of Bideford	Manure from Bristol to Bude	Bude Harbour
1876 December	*Hawk* of Bideford	Limestone from Aberthaw to Bude	Entrance to Bude Haven
1878 March	*Friendship* of Bideford	Patent manure from Ilfracombe to Bude	N side of Bude Haven
1879 December	*Catherine* of Fowey	Coal from Newport Mon to Bude	Entrance to Bude Haven

Date	Vessel	Cargo	Where wrecked
*1880 March	*Bee* of Bideford	Culm from Swansea to Bude	N side of Bude Haven
1880 December	*Stuckley* of Padstow	Coal from Mumbles to Bude	Bude breakwater
1884 November	*Joseph and Thomas* of Bideford	Coal from Port Talbot to Bude	Bude Haven
1885 October	*Hawk* of Bideford	Coal from Newport Mon to Bude	Entrance to Bude Haven
1886 January	*Eling* of Lime	Sandstone and wool from Plymouth to Bude	Entrance to Bude Haven
1886 February	*Joseph and Thomas* of Bideford	Coal from Port Talbot to Bude	Pier Head, Bude Haven
1886 November	*Lady of the Lake* of Bideford	Coal from Port Talbot to Bude	Entrance to Bude Haven
1887 January	*Sir T. D. Acland* of Bideford	Coal from Newport Mon to Bude	Bude Haven
1887 October	*Endeavour* of Bideford	Coal from Lidney to Bude	Entrance to Bude Haven
1888 February	*Alford* of Bude	Timber from Plymouth to Bude	N side of Bude Harbour
1888 February	*Elizabeth* of Bideford	Manure from Plymouth to Bude	N side of Bude Haven
1888 November	*Ant* of Bideford	Culm from Swansea to Bude	Bude
1888 December	*Joseph and Thomas* of Bideford	Slates from Port Gaverne to Bude	Rocks N of Bude Haven
1891 March	*Elizabeth* of Bideford	Ballast from Bude to Newport Mon	Inside Bude breakwater
1893 November	*Ceres* of Padstow	Coal from Newport Mon to Bude	Summerleaze Cliff, Bude
1895 August	*Alford* of Bideford	Coal from Garston to Bude	Chapel Rock, entrance to Bude Haven
1895 August	*Hawk* of Bideford	Coal from Lidney to Bude	Bude Haven
1896 March	*Katie* of Bude	Bricks from Bridgewater to Bude	Entrance to Bude Haven
1896 October	*Annie Davey* of Liverpool	General goods from Liverpool to Bude	Entrance to Bude Harbour
1897 February	*Brackley* of Liverpool	Timber from Swansea to Bude	N entrance to Bude Harbour
*1897 February	*Jersey Packet* of Guernsey	Manure from Ipswich to Bude	N entrance to Bude Harbour
*1897 August	*Hawk* of Bideford	Scrap iron from Bude to Newport Mon	Entrance to Bude Harbour
1897 September	*Eva Potter* of Hull	Railway material from Middlesboro' to Bude	Bude sands
1897 October	*Spirit* of Cowes	Coal from Newport Mon to Bude	Bude
1897 November	*Esther* of Plymouth	Coal from Newport Mon to Bude	Bude sands

Page 157 (*right*) A typical boundstone used to mark the extent of the canal company's property; (*below*) A stretch of the former canal near Parnacott showing the reed-filled bed

Page 158 (*above*) Site of the former canal basin and an old warehouse building at Blagdonmoor wharf, terminal of the Holsworthy branch; (*below*) A stretch of the maintained feeder arm showing one of the numerous bridges which cross it

Date	*Vessel*	*Cargo*	*Where wrecked*
1898 August	*Purveyor* of Southampton	Bricks from Bridgewater to Bude	N side of Bude sands
1898 September	*Lady of the Lake* of Bideford	Coal from Newport Mon to Bude	East sands, Bude
1898 November	*Wild Pigeon* of Padstow	Coal from Lidney to Bude	Bude sands
1899 January	*Elizabeth* of Bideford	Coal from Newport Mon to Bude	Bude
1899 April	*Annie Davey* of Bideford	Manure from Ipswich to Bude	Bude sands
1899 May	*Teazer* of Padstow	Manure from Plymouth to Bude	Bude sands
1899 May	*Ceres* of Padstow	Ballast from Bude to Falmouth	Rocks on W side of Bude Harbour

* Total wreck.

APPENDIX 12

Hobbacott Down—The Great Incline

THE following extracts about the Hobbacott Incline were taken from various sources; the Journal kept by the Inspector from 19 October 1820 to 11 January 1822, the Monthly Reports of the Canal and Works from John Honey, Clerk to the Company at Bude, letters from John Honey to the Company chairman and others, Minutes of the Exchequer Loan Commissioners, Resolutions and Reports of the Committee of Management, Reports of the Visiting Committee and various letters.

1820. October 26th

Took the line from Anderton downwards to Bude, party of men working on Obbacott Downs but the works below all deserted on account of the heavy rains.

1820. November 10th

. . . Parties working on Obbacott Downs House on that part of the work on progress. The Culvert at the foot of the Great Incline finished. . . .

1820. November 24th

. . . The Shafts on Obbacott Downs proceeding but slowly owing to the water. One shaft is about 50 and the other 40 feet deep. . . .

1820. December 27th

. . . One of the Whims at Obbacott at work and one shaft expected to be clear of water this night. Plane partly forward and each shaft sunk about 70 feet and the adit driven about 60 feet from the foot of the Plane to Cann Orchard, the ground open but no part can be said to be completed.

1821. January 1st

. . . Whims at Obbacott both at work and both shafts expected to be clear of water this evening.

Extracts from the Journal

1824. September 9th

. . . From the Trade that has hitherto gone along the Canal, no correct criterion can be formed of what may take place; the demand is not regular, but brisk and slack at intervals, as the farmers find employment on their farms; so that nothing less than twelve months is sufficient to form an idea of the probable Trade.—Some days 50 Boats will pass up the Planes, but on an average it does not exceed 35—I could not calculate on the Engine rising more than 60 Boats each working day. Not having seen both Buckets in a working state, I cannot with confidence state the number of Boats that might be raised by them each day; but from the observations I have made from one Bucket working, I am inclined to think that 100 Boats might be raised—400 tons at 5 working days—2000 tons per week and at 1/9d per Ton (the average of the Tolls already along the Canal) will be £175 per week or £8,400 per year. This I apprehend will be the greatest extent that the Machinery on the Planes can perform. . . .

Letter from Alexander Easton to Thomas Telford

1826. November 30th

The upgoing Rail Road is in very good repair it having been recently laid with new metals on an improved plan which answers extremely well. The downgoing road is in a very bad state, being so broken in short lengths; The Machinery has for a long time been worked with success and free from accidents, and is at present in good repair, there is a new bucket ready in case an accident should happen. The Masonry and Timber frames are in good order and preservation with the exception of a few rotten places in the frame work, but which at present has not the least appearance of giving way. The Steam Engine which has not been worked for some time is in good order and ready for immediate use.

Monthly report from John Honey

1826. December 27th

December 14th. Information arrived at this office in the evening that both Chains of one of the Buckets at Hobbacott Down had broken and the Bucket had fallen from half way the pit to the bottom. I went next morning and examined the same, when I found that the injury done was not so much as might have been expected. The Bucket can be repaired on the spot, and the earliest opportunity will be taken to do it, but before this can be done the new Bucket which was in store must be used, and which is now prepared and in use. The framing at the bottom of the pit was broken to pieces, but which is now replaced with new and placed in the pit in a very substantial manner, in such a way as I trust will answer well its intended purpose. The Steam Engine has been worked ever since the accident happened.

Letter from John Honey to J. H. Stabback Esq

1827. January 4th

. . . The Great Inclined Plane has given much trouble since the last Report, by the breaking of the Chains of one of the Buckets. The Bucket has been much damaged but will admit of being repaired, and the spare Bucket has been got in and is now in Work. The Steam Engine has been in use the day previous to the Accident, as the Timber framing which receives the Bucket had been damaged and was under repair when the Bucket fell and destroyed it altogether. This Machinery will require much care and attention, and I fear the expense of repairs will be considerable.

Letter from George Call to the Committee of Management

1827. January 14th

I have received from Mr Honey . . . Information that the Chains at the Great Incline have again parted, that the Bucket when about three parts down fell and has been materially injured. The Steam Engine which has been put to work also requires some trifling repairs of Timber of which we have none under hand—The Committee are not aware of the Difficulty of getting Materials at a moments notice and I hope they will reconsider the question of having some in store—As for the Bucket I would recommend serious consideration of the propriety of increasing the Strength of the Suspension Chains—these having now given way repeatedly. Mr Honey and Geake the Mechanic at that Plane have examined every link when the last casualty occurred and cannot account for it but by want of substance in the Chain itself. . . .

No time should be lost in determining whether new Chains are to be ordered or the old ones repaired as the expense of working the Engine will be considerable.

It would save much time and fuel in getting the Steam Engine to work and keeping up the fire if the furnace were inclosed in a Shed which might be erected at a trifling expense, the loss of Heat working in the Air is considerable, and the Iron exposed outside the Engine House must suffer.

Letter from G. C. Call Esq to J. H. Stabback Esq

1827. January 31st

The Bucket which last fell to the bottom of the pit on the 12th Inst. is nearly repaired, and I hope will be put on work on Monday or Tuesday next. The Chains of both Buckets as also the Main Chain of the Plane have been examined in the most accurate manner, and every link that had the least appearance of being defective has been taken out and replaced with new Rivet links made from the best Iron and I trust they will continue in work without accidents for a longer period than they have done for some time past. During this last week I have examined the flues around the Steam Engine Boiler which I found necessary to have cleaned out, they being very foul. In examining the flues I found the Boiler had been much injured from its being so exposed to the weather, the Water passing between the bricks and the boiler has caused that part of the boiler where the flues are carried to rust in such a manner that shells of iron nearly a quarter of an inch in thickness have fallen off from it, this certainly gives a sufficient proof that the Boiler must diminish in strength and shows the necessity of covering the same to prevent further injury as well as a great saving of fuel.

Monthly report from John Honey

1827. March 6th

This inclined Plane is in good repair, and the Buckets which were injured are now put in perfect repair, one of them suspended in the pit and in work, and the other near the spot in readiness to be put in work in case an accident should happen. The Machinery is in very good repair and works extremely well.

Monthly report from John Honey

1827. April 2nd

The ascending Road is in good repair but the descending one is similar to that at Marhamchurch, the Metals being much broken. The Machinery has worked extremely well ever since my last report, until 27th March last, when in taking up a loaded Boat the Ring Bolt which hangs it on the Main Chain broke out a few yards up from the bottom, no damage was done to the Boat, but in letting down the heavy Brakes to stop the progress of the Bucket from descending in the pit caused the end framing on which the Drum shaft runs and where the Brake takes its pressure to give way, the same has been temporary repaired so as not to stop the Trade, but it will be necessary as soon as possible to repair it more effectual.

About 2 o'clock this day a Report was brought from the Great Inclined Plane that one of the Bucket Chains had again broken, the Bucket had descended about three quarters of the depth of the pit. The Bucket is now hanging on by one Chain, which Chain with the weight of the Bucket in falling over on its side, parted the side of the Bucket from top to bottom, the Bucket can again be repaired on the spot and will take nearly a week to complete it. The Engine was immediately put to work so as not to impede the Trade.

Monthly report from John Honey

1827. May 2nd

. . . The Machinery and Canal Banks have failed repeatedly and the consequent stoppage of the Trade has materially affected the Revenue. . . . Under the first head the Chains have parted, Buckets have been broken, Metals on the several inclines have given way, (and we have none in store to replace them). This occasions damage to the Boats passing, injury to the Machinery by irregular movements, and the expense of Coals working the Engine whilst the other part is under repair has been considerable.

Report from George Call to the Committee in Exeter

1827. June 4th

May 5th. Both Bucket Chains in the West pit at Hobbacott Down broke and the Bucket went to the bottom and injured it so much, that the expense of repairing will not be less than from £20 to £25.

May 10th. The Main Chain of the Inclined Plane at Hobbacott broke in taking up a Boat of Mr King's laden with Sand, the Eye in the Boat which the Scotch is attached to in ascending over the Planes gave way which was the cause of the Boat descending, this Boat at the bottom in descending came in contact with two others, and injured them very considerably, as the Company are in want of Boats I beg leave to suggest the propriety of purchasing these damaged Boats, the expense of which will be from £25 to £30, when they can be repaired and answer for a long time for the Company's use.

May 16th. The Main Chain at Hobbacott Down Incline Plane parted but no further damage was done.

Letter from John Honey to J. H. Stabback Esq

1827. July 2nd

The upgoing Road on this Plane is in perfect repair, the down going road is in bad repair the Metals being very much broken. The Bucket Machinery is in a perfect state of repair and works extremely well—59 Boats were taken over this Plane on the 25th ult. in 7 hours.

The Main Chain at Hobbacott has broken twice this last month.

Report from John Honey

1830. October. Following a breakdown at Hobbacott new wheels were ordered from Wales. There was a long delay in delivery. The use of the Steam Engine was restricted owing to the price of coal, and trade consequently suffered.

1830. November 23rd

The Wheels are not as yet arrived to Bude from Newport.

There are a few setts of Boats which the Traders have applied to have taken down over Hobbicott Plane to lay up, I have refused to take them down, as there is none loaded to come up against them, and the expense of Coals in getting up the Steam in the Engine would be all for no gain whatever, what am I to do in this case.

1831. January 11th

I beg to inform you that the Wheels arrived yesterday at Bude safe from Newport in the Sloop Ceres of Dartmouth, Capt. Knowles, the remainder part of his cargo is coals, I hope to get them to Hobbicott Down by Saturday night next. The Vessel from want of water being low tides cannot come in the Basin before tomorrow (Wednesday evening).

1831. January 20th

The large wheels being deposited on the spot where they are to be fixed . . . I beg to inform you that every thing is ready for hanging them.

1831. February 11th

I am extremely sorry to inform you that the Boiler of the Steam Engine is rusted through in many places, which prevents us from working at present in any way at this place. I had some suspicion that all was not right in the Engine Boiler which led me to make a thorough examination, and in examining it, found that between the Wall and that part of the Boiler next to it, was completely eaten out with rust, and in two places found holes completely through it three inches square. It is to me miraculous that the Boiler had not bursted and perhaps killed the person who worked it, luckily no misfortune had happened. I can repair it here, which will be immediately attended to as soon as the Bucket Machinery is ready which I hope will be nearly completed by next week. The Engine Boiler is injured by means of the water dropping from the Engine House and resting between the side of the Boiler and the Wall, it was all exposed, before I covered it in about 4 years since. I have no doubt after I get it mended, and the flues made in a proper manner, and the valves made tight and the pumps put in order but that I shall be enabled to do as much work with 15 bushels of Coals, as what we now use 25 bushels for.

Letters from John Honey to J. Blackmore Esq

1831. August 12th

The Beam of the Chain Wheel at Hobbicott Down being so much decayed as to render it almost immediately necessary to put in a new one. . . . The Wood work of the Drums and Brakes at Hobbicott Down are very much decayed, there is no wood in hand to repair it; something must be done soon otherwise the whole will fall a wreck together.

1831. August 13th

About half past twelve o'clock this day a man named John Wilkey a boatman in Mr Kings employ in the act of taking down some empty boats from the top of Hobbacott Plane, hooked on three and put the shackle on the Main Chain in the regular way; immediately as the two foremost boats came out over the top of the Plane, this man Wilkey, who was in the foremost boat, was seen by William Dyer a mason working for the Company removing or handling the shackle; and by this means the Boats escaped from their fastening and run from Top to bottom of the Plane; on examining the foremost boat, I found the V which the Main Chain falls into also broken, as they came to the bottom the foremost boat ran immediately into the centre of the Work and Broke the Shaft of the Chain Wheel and several of the Wood arms and has considerably damaged the framing. . . . The damage done will occasion the Trade to stop for 10 days or a fortnight.

1831. September 8th

It will be my greatest pride to put and see the whole of the Machinery in good order and the Trade in a flourishing state, and nothing shall ever be wanting on my part to effect it, but from the present appearance of repairs which in justice ought to be done to the Machinery at Marhamchurch, Hobbicott Down, Vealand and Merrifield Planes and the expense always attending that at Hobbacott, that I have some reason to fear from the extensive and complicated Works at the latter place, it will always be a Mill Stone on the Companys neck. I well know it is now no time for reflection but had this Plane been fortunately made in two, in different direction and worked by two water wheels, almost all the former enormous expense, and that which in future must be, would have been saved.

Letters from John Honey to J. W. Crabb Esq

1832. February 4th

I am sorry to be under the necessity of thus troubling you with the unpleasant information respecting Hobbacott Down Machinery, which was very much damaged on the 1st ins. by the Main Chain breaking.

Letter from John Honey to Wm Holden

1834. June 19th

The Memorial of the Bude Harbour and Canal Company sheweth . . .

That subsequent and expensive Casualties have occured to the Machinery generally employed on the Canal and more especially to that whereby the transit of the Boats is effected at Hobbacott Down.

That within the last two years it has been found necessary to erect new Machinery at that place in consequence of which the Trade has been greatly impeded and a considerable expense incurred which has absorbed all the receipts of the Company.

That the experience of your Memorialists is in favour of some modification or change in the Machinery of the inclined Plane and that herein they are prepared to attend to the suggestions of the Secretary of your Hon. Board.

. . . signed William Pratt Call. Chairman of the Committee of Management to the E.L.C.

The Secretary reported that he had inspected the Works of the Bude Harbour and Canal Company.

. . . That generally speaking, the Canal and Planes have been practically completed, and although requiring repairs, remain with one exception in tolerable working order.

That the principal impediment to success has been the Plane at Hobbacott Down. This plane is made to rise 1 in 4, 900 feet in length, the perpendicular height being 225 feet.

That instead of a simple and efficient contrivance for carrying up the Boats to the higher level, the complex method of movement by perpendicular Water Buckets with multyplying Wheels etc invented by Fulton the late famous American Engineer, and published in his pamphlet, has been adopted.

That this complex Machinery has been subjected to repeated accidents of a serious nature, leading to great expense in repairs, causing impediment in the Traffic, want of confidence in the Traders, and consequent loss of Revenue, so that the Canal and Works have got into a state of dilapidation without remedy unless the course of management is very materially altered.

That there is a prospect of an increasing and remunerative traffic if the chief difficulty can be removed.

That more than 80 loaded Boats were stopped at the great inclined Plane on the days of the Secretary's inspection, that he conversed with several of the Traders and found one uniform opinion that the Traffic would greatly increase and the Agriculturalists and Public be greatly accomodated if proper facility of transit were given.

Report of the Secretary to the Exchequer Loan Commissioners

1835. May 14th

It appears . . . that the Committee are desirous to contract with Mr James Green, the Engineer, to repair and improve the Machinery at Hobbacott Down at a cost of £300, and to work and keep the same in repair for one year for the further sum of £260 to be paid in monthly instalments, giving him sole control of the Premises and the Works at Hobbacott Down.

That by the report of Mr Rendell, in his opinion the principle on which the Bucket Machine at Hobbacott Down is constructed, is not such as can be depended on for carrying on the Trade.

1835. June 11th

The Secretary reported that it appears that the present party in power in the Management of that undertaking have determined not to adopt the powerful Steam Engine recommended by their former Committee to be erected and employed for raising the sand up the Hobbacott Plane instead of the Bucket Machinery and Small Steam

Engine; but have contracted with James Green to repair and manage the Bucket Machinery at the expense under mentioned viz.

For repairing the Bucket Machinery	£300
For one years current expenses to be paid to Mr Green	£260
For the small steam engine	£100
	£660

... As Mr Green's intention is only to use the small steam engine during periods when the Bucket Machinery is rendered unserviceable by accident, his proposal to incur expense for repairing the small steam engine at the same time with the Bucket Machinery shews no confidence even on his own part in the stability of his Bucket Machinery.

Minutes of the ELC

1837. May 25th

In reply to yours of yesterday, I believe the engine is 16 hp. The Boiler at present in use is nearly worn out, and it is not trustworthy, we want a new one on the best principles.

Letter from J. Waddington to Messrs Harvey & Co, Hayle Foundry

1837. May 25th

... The Engineer was directed to employ Charles Brown as Conduiter of the Plane at the rate of 3/6 per Diem a House to live in and £4 at the end of the Year provided no accident happened to the Machinery by his Neglect, also that Thomas Shepheard be employed as his assistant at the rate of 3/0 per diem and £2 at the end of the Year provided no Accident occur through his Neglect.

That the Engineer do forthwith clear out the present Blacksmith Shop for a Dwelling place for Thomas Shepheard and that the old Smiths Shop on the west side of the Plane be forthwith used by the Blacksmith.

The Committee most positively desire the Engineer on no account whatever to allow the Conduiter or his Assistant at Hobbacott Plane ever to be removed from thence to assist on any other part of the Line there being always sufficient Work in oiling and examining the Machinery when the Plane is not at Work. ... It would be as well if the persons passing Boats over the Plane could be prevented holding conversation with the Conduiter or his Assistant whilst passing the Boats up or down and on no account whatever is any person either Proprietor or otherwise to be allowed to hook a Boat either down or up the Plane. All access by strangers to the Plane on Sundays should be prevented.

Resolutions of the Committee of Management

1839. January 15th

That a sufficient quantity of Oak plank of the thickness required for the completing the Drums at Hobbacott be immediately procured by the Engineer.

1839. March 19th

That the Engineer do stop the Canal to Refix the Drums at Hobbacott at his discretion.

1839. June 25th

That the Engineer be directed to forward to the Committee a plan of the intended new Buckets at Hobbacott Plane.

1840. February 18th

That the report of the Engineer on the Improvements made in the Valves of the Buckets is satisfactory.

That the Engineer . . . do make such alterations to the Governor Pumps as he may think necessary, forwarding to the Committee the plan of his improvements and also his plan for New Rollers for Hobbacott Plane.

1840. May 27th

That a Chain be immediately ordered for Hobbacott Plane.

Resolutions of the Committee of Management

1873. August 27th

Buckett chain broke in North Pit, a rivit link.

1873. November 21st

Chain fell into Pit today. Never saw anything like it before—so wild.

Extracts from the diary of James Sleeman

1877. September 4th

All the works were in good order. The only thing to report on is one of the cables of the new chain has had to be replaced, this should not be considering the heavy sum expended on it so recently. The

Steam Engine and Boiler here we strongly urge on the Committee the necessity of selling before the iron becomes quite useless. Any attempt at working the Plane with it would be attended with disastrous consequences and loss to the Company.

1878. July 20th

One half of a new chain is required as the old part is in very bad order and weak. The whole of the Timber on the south side requires being renewed it having stood 40 years. The strain upon the frame is very great and should the chain break and a boat slip back the damage would be very serious.

Reports of Visiting Committee

1878. July 3rd

Mr Sullivan reported that 1000 feet of new chain was required at Hobbacott.

1878. September 5th

Hobbacott Chain. Harries of Bristol tender accepted 17/6 per cwt.

Minutes of Committee of Management

1879. July 8th

The Plane appears to be in good working order. All the repairs to the woodwork at the head of the Plane are not completed. . . . Stonework at foot of Plane in want of repair.

Report of Visiting Committee

1879. May 7th

Half a new chain has been purchased for Hobbacott.

1880. March 31st

Improvements—Timbering and plating one of the drums at Hobbacott.

Annual Report of Committee of Management

1880. August 16th

Mr Sullivan is about to entirely renew the timber work at the top of this Plane, the wood was all prepared and ready when the Committee was there, and the work was to proceed on the 17th August when the traffic was to be stopped.

Report of Visiting Committee

1881. May 4th

The framing at Hobbacott Plane has been renewed.

Annual report of the Committee of Management

1882. August 3rd

Brown of Hobbacott is thought to be too old to continue in his present post and it is recommended that he gives it up.

1882. October 8th

Charles Brown Sen. is requested to leave and give up possession of his cottage. Charles Brown to be allowed £6 per annum for the present in consideration of his advanced age and long service.

1882. December 7th

An attempt has been made to damage the machinery and metals at Hobbacott. A reward of £20 offered for information.

Minutes of Committee of Management

1883. May

A large amount of repairs has been done to the framing and buckets at Hobbacott Down Plane.

1884. May 7th

Framing and Buckets at Hobbacott Down Plane thoroughly repaired.

1885. May 6th

Buckets at Hobbacott Down thoroughly repaired.

1886. May 5th

At Hobbacott a new top and liner and brake have been fixed and the small drum timbered.

Extracts from Engineer's reports

1886. August 13th

The Committee regret to find that a considerable sum will at once have to be spent here to make the working of the Plane safe; a great deal of woodwork being rotten and two new stays being required at the bottom. This the Engineer proposes doing when the Canal is stopped at Harvest time.

Report of Visiting Committee

1887. May 4th

At Hobbacott very considerable repairs have been done to the framing, stays and bottom sills, and a part of the timber road had been renewed.

Engineer's report

1887. July 30th

The repairs to woodwork ordered last year appear to have been satisfactorally carried out. When the Committee visited the place the Carpenters were busy repairing the bottom of one of the Buckets.

Report of Visiting Committee

1888. May 2nd

Very important repairs have been done to the bucket at Hobbacott.

Engineer's report

1888. July 14th

At Hobbacott, Sleeman's House requires repainting badly. Sleeman is willing to do the work when not on duty if the Company will provide the paint.

1890. July 8th

At Hobbacott some of the Cross timbers are rotten and (if the Canal were to be kept open) should be renewed. In view of its abandonment however the Committee consider that some temporary struts will make the place secure for a time.

Report of Visiting Committee

1893. January 19th

Sullivan is ordered to cover in the Hobbacott Wells with concrete.

Minutes of Committee of Management

APPENDIX 13

Inclined Planes: an Assessment of Fulton's Contribution

UNDOUBTEDLY, the most remarkable aspect of the Bude Canal's construction was the system of inclined planes which it included, the ingenuity of their mechanics and the fact that they were no less than six in number—more than on any other canal in Britain either before or since. In this connection several questions arise. By what steps did the evolution of the planes' eventually agreed-on plans progress, and by what and by whom were they influenced? And particularly, in this respect, the extent of the actual original and genuine contribution made by Robert Fulton. How much of a success or failure were the Bude Canal's inclined planes and how far, if at all, did they influence canal construction elsewhere? Some answers it is possible to make with certainty, others must remain unresolved, as enigmatic as the character of Robert Fulton himself.

While tendencies exist for Fulton to be accredited with a major share of the ideas for the design of the Bude Canal planes it is pertinent to consider what already existed in this field when Fulton came upon the scene. There had, of course, been the earlier proposals for this particular canal which had not taken shape, and there were a number of previous canals where a system of inclined planes had been devised.

The idea of inclined planes for negotiating changes of level in waterways is of much earlier origin than the times generally regarded as the 'canal age'. While the Greeks probably used rollers to aid the transfer of ships over the isthmus of Corinth as early as the sixth century BC, the first known example of actual but primitive inclined planes is provided by the Chinese, who hauled vessels up and down

Page 175 (*above*) Moreton bridge over the canal feeder from Tamar Lake, showing the original cast-iron arches; (*below*) The entrance to the wheelpit at the top of the former Tamerton inclined plane

Page 176 (*Above*) The slope of the former Werrington or Bridgetown inclined plane as it crosses the public highway; (*below*) This building at Druxton wharf is typical of the canal storehouses

sloping planes between upper and lower levels of canals by means of manually-operated capstans sited on either side of the incline at the top, a system which, surviving up to the present century, may have been in existence for nearly two thousand years.

On the continent there were inclines on canals from medieval times. In Flanders they were in existence by the twelfth century and one there was worked by a waterwheel up to the seventeenth century. There were four, worked by man or horse-gin, on the Ypres–Nieuport Canal and in the seventeenth century there were inclined planes in Holland, operating on rollers worked by capstans, at Overtoon, near Amsterdam. Designed for a different primary purpose was an incline existing in the fifteenth century in Italy, which was made to prevent fresh water and silt from the River Brenta flowing into the Venice lagoon; vessels were carried over it in sledges transported on rail tracks, powered by a horse whim.

It was an Italian engineer, Davis Ducart, who built the earliest canal inclined plane for carrying boats in the British Isles, in Ireland on the Coalisland Canal in County Tyrone, during the early 1770s. Three planes were planned and it appears that they were at first constructed with timber ramps fitted with rollers, but difficulties arose over providing suitable power for hauling up the small 1 ton boats. On Smeaton's suggestions the arrangements were changed. Double ramps were constructed so that the boats could be counter-balanced, the boats were increased in size to 4ft 6in wide, 10ft long and 2ft 6in high, and horse gins were provided for hauling them out of the upper pounds. But even this system did not work very satisfactorily and Ducart substituted parallel railways on which the boats could ascend and descend, each resting on a cradle fitted with four wheels. The canal with its three 'hurries' or 'dry wherries', as the planes were called locally, was completed in 1777 but its existence was short-lived. The gradient of the planes proved too steep for successful operation by counterbalance and by 1787 these and the canal itself were inoperable, the whole scheme having been an impracticable proposition.

The first canal inclined plane constructed in England was on the Ketley Canal in east Shropshire, built by William Reynolds and completed in 1788. Although Reynolds could have been influenced in his design for the plane by earlier ideas elsewhere—the 'hurries' of the Coalisland Canal, or chutes or slides such as were used in Cornwall in conjunction with Parnall's Canal and the St Columb, or even by Edmund Leach's proposals of 1774 for the Bude Canal—the possibility that Reynolds knew about these is considered unlikely, and it was probably simply his own ingenuity, and perhaps the thought of the ordinary horse tramways which were often constructed in conjunction with canals and frequently made use of inclined planes, which provided the inspiration. The Ketley was a tub-boat canal and records of the plane's vertical height vary from 67–73ft. Almost all the traffic—coal and ironstone from Oakengates to Reynolds' Ketley works—was downwards and the loaded 8 ton boats by their weight themselves provided the motive power. They were carried over the incline in cradles which had two small wheels at the higher level and two large ones at the lower level, so that the boats rested on a level platform. Ascent and descent were controlled by a brake operating on the axle of an overhead drum set at the top of the plane, over which was wound a rope attached to the cradles. The cradles travelled on two parallel sets of rail tracks and at the top of the plane passed into a lock which connected with the upper pound of the canal.

Following immediately after the completion of the Ketley Canal came the construction, between 1788–92, of the Shropshire Canal. Here again in this hilly countryside changes of level were involved which were overcome by three inclined planes, the Wrockwardine or Donnington Wood, of 120ft vertical descent, the Windmill (126ft) and the Hay or Coalport (207ft). These worked on the same general principle as the Ketley plane except that instead of the lock connection these planes each had a short reverse slope at the top, which obviated the use of water for boat transfer. Also, steam power was employed. From Brierly Hill, the terminus of the canal's westward branch, connection was made with the Coalbrookdale works; at first this was by two vertical shafts at the canal side, through which coal and iron were lowered and limestone lifted, with tunnels leading to the works, but in 1794 these were replaced by a counter-balanced inclined plane carrying waggons, loads being transferred between them and the boats. (In about 1790 on the Donnington Wood Canal, with which the Shropshire Canal connected in the north, a similar arrangement of interchange between levels by means of vertical shafts had been replaced by an inclined plane which may have carried boats, as, from 1797, did the underground inclined plane on the Duke of Bridgewater's colliery canal at Worsley, near Manchester.) Soon after the Shropshire Canal was completed the Shrewsbury Canal was being built, and was opened in 1797, this too included an inclined plane, the one at Trench, of 75ft vertical height and worked, in its early years, on the same system as those of the Shropshire.

The innovation of the boat-carrying inclined planes of the Shropshire Canal was of much interest to visitors, who came from near and far to view the system in operation. Among the visitors was Robert Fulton.

Fulton, an American descendant of emigrants from Ireland, had arrived in England probably at the beginning of 1787, in the hopes of developing his artistic talents and undoubtedly to seek a living if not a fortune. For a time he worked under an artist in London and in 1791 portraits painted by him were exhibited at the Royal Academy. Further exhibitions followed in the same year, and Viscount Courtenay commissioned Fulton to paint his portrait, for which Fulton travelled from London to stay at Powderham Castle, Lord Courtenay's seat near Exeter, a trip which probably marked his first connection with the West Country. This commission led to others from influential people and Fulton's future appeared hopeful. For this reason it is surprising that at this stage, when in his thirties, Fulton decided to abandon his artistic career in favour of engineering, his other bent. He seems to have made the decision while still in Devon and his first mechanical accomplishment appears to have been a mill for sawing marble which was put into use near Torbay, a model of which earned him the silver medal of the Society for the Encouragement of Arts, Commerce and Manufactures.

Fulton's thoughts then turned to canal construction and the movement of vessels by steam power, subjects on which he developed correspondence with the third Earl Stanhope who, with his own inventive genius (he was the inventor of the Stanhope levers in printing presses and the Stanhope lenses in optics) had already been giving much attention to these particular matters. It was unfortunate for Fulton that his idea for preponderating cisterns of water for drawing boats up inclined planes had already been put forward by Edmund Leach; nevertheless Lord Stanhope did not discourage him. On 8 May 1794 Fulton took out his only patent, entitled 'A machine or Engine for conveying Boats and Vessels and their Cargoes to and from the different levels in and upon Canals, without the Assistance of Locks or the other Means now known and used for that purpose'. The specification included use of the previously suggested water cistern system of counterbalance, either in conjunction with an inclined plane or a vertical shaft lift—more improved methods of which were already in use—but brought in a new suggestion (possibly inspired by Lord Stanhope's previous idea for carrying boats up rail inclines suspended between pairs of wheels), for wheeled carriages to enable boats to be hauled up the reverse slope at the top of the incline and into the cistern vehicles. From this idea probably derived the obvious improvement, to put wheels on the boats and so eliminate carriage and cistern alike. A friction brake was to be used for checking the speed of descent.

At this time, and partly no doubt due to the expense of obtaining his patent, Fulton's funds were running low. He continued his travels up and down the country, viewing much of the considerable amount of canal building then in progress. While in Manchester he struck up a friendship with Robert Owen, a cotton manufacturer of about his own age whose financial assistance he enlisted in the development of further ideas. One of these was a machine for digging out canal beds, another idea which he communicated to Lord Stanhope, and which he hoped to use on the Gloucester & Berkeley Ship Canal, but one which did not get much farther than the drawing board. Concurrently with formulating his ideas for his digging machine Fulton, in a brief partnership with Charles McNiven, in October 1794 accepted a contract for cutting part of the Peak Forest Canal. Probably because of the time he was spending developing his machine and his preoccupation with the Gloucester & Berkeley (and the time McNiven was spending elsewhere) it was reported by Benjamin Outram, the Peak Forest's consulting engineer, that the partners were not proceeding with the cutting for which they were contracted. Either at this time or a little earlier—perhaps both—Fulton had visited east Shropshire where he had seen the inclined planes of the Ketley and Shropshire Canals; he mentions them in his later book and in a letter to Lord Stanhope, and in 1795 he suggested the possibility of using tub-boats and an inclined plane on the Peak Forest Canal, following which Outram and two others were sent to view the system at Coalbrookdale. They reported that small boats and inclines would be practical and save money and Fulton was requested to have printed at the company's expense his ideas and the comparative advantages of using 20 ton, 5 ton and 2 ton boats and railways. He was rewarded by the company with 100 guineas for his ideas for the canal, but appears to have severed his connection with the Peak Forest very shortly afterwards, without his ideas having been put into effect.

Fulton's book *A Treatise on the Improvement of Canal Navigation* was published early in 1796 and was probably being written at the time he was employed by the Peak Forest company. It is possible that the company financed the book's production, and, if so, it was an indiscretion that he omitted to mention it. In the book, which attracted considerable attention and influenced later developments not only in this country but also in France and America, Fulton advocated small canals and small boats. He had come to the conclusion that transferring boats on to cradles at inclined planes involved a waste of time, and as an alternative he considered the merits of constructing the plane of rollers or alternatively of providing the boats with wheels to run directly on the plane; he decided in favour of the latter, bearing in mind that if a set of wheels went wrong one boat only need be withdrawn for repairs but if rollers were out of commission the whole trade would be stopped. Also, as has already been described (p 24), Fulton proposed the method of the tub or bucket of water in the well as a means of providing power, as well as the alternative use of waterwheels. The bucket-in-the-well method was an entirely new departure, the idea for which, on the eventual Bude Canal, as for the wheels on the boats, Fulton must be given the full credit. One wonders at the stages by which the well and bucket method was formulated in his mind: perhaps it was only a few logical steps on from his system of counterbalance in a vertical shaft lift, specified in his 1794 patent, or could it be that some thought of the vertical shafts connecting the Shropshire Canal with the Coalbrookdale works, through which material was lowered and raised, which were in use until 1794, had lingered on in his mind?

Following the publication of his book Fulton was asked to report on a proposed Helston canal across south-west Cornwall and planned it to include six inclined planes, but after more correspondence with Lord Stanhope—who at this stage had further ideas for canal lifting in the form of a 'pendanter', another type of vertically-operating, counterbalanced boat-carrier—he departed from the immediate scene and turned his attentions to his work on submarines, torpedoes and steamships, which he pursued in France and after returning to America. Ideas for the Bude Canal were suspended during the years of war and by the time they were revived, in 1817, Fulton had been dead two years.

Then it was up to James Green, backed now by the fourth Earl Stanhope, to solve in practical terms the changes of level which the hilly countryside demanded and to make the system work. His decision in favour of small boats working on a small canal with inclined planes is believed to have been influenced by his knowledge of the Shropshire canals, and yet he adopted the ideas for wheeled boats and the methods for the use of water power for lifting that Fulton had suggested.

Undoubtedly the use of inclined planes was the best solution to the problem of the changes of level in this particular situation, due to their cheapness compared with locks and their greater economy of water. And, with the predominantly in-going traffic, much of it intermittent, the use of water power was more practical than relying on counterbalance. But the five waterwheel-operated planes of the Bude Canal, all together, probably caused less trouble than did the single huge Hobbacott plane with its bucket-in-the-well method of lifting. It was not so much that the principles of design were wrong, for at times the system worked surprisingly well, but so often the materials could not stand up to the strain, and when there were breakages of chains and other metal parts repairs could not be effected as quickly as in Shropshire, where foundries were close at hand. The frequent failures at the Hobbacott plane were indeed the crux of much of the Bude Canal company's financial difficulties.

But Green must have had confidence in the inclined plane system, for he included one in his construction of the Torrington Canal (opened in 1827), powered by a waterwheel, and another at Wellisford on the Grand Western which he completed in 1838. The latter was designed on the same principle as the incline at Hobbacott—surprisingly, in view of the difficulties that were there by then being experienced—but differed in that the boats were carried over it in cradles; the water power provided by the bucket-in-the-well system proved at Wellisford to be inadequate for lifting the boats of the Grand Western Canal—double the weight of those of the Bude—and a steam engine had to be substituted.

Another West Country canal which employed inclined planes was the Chard, completed in 1842. Here there were four planes; three of them were double track one of which, according to information only recently available, was powered by a waterwheel sited above the plane at the foot (the other two probably being worked similarly), and the fourth plane was of single track, powered by a water turbine. James Green also designed three inclined planes for the Kidwelly & Llanelly Canal in South Wales, only two of which were completed, during the 1830s; apparently they worked on a system of counterbalance, aided in some way by hydraulic pumps.

In later days further inclined planes have been constructed, but for bigger boats; one was constructed in 1857 on the Monkland Canal in Glasgow and another in 1900 in Leicestershire for by-passing the Foxton locks on the Grand Junction. In each case the boats travelled on caissons and were counterbalanced, aided by steam power, but their working lives were of limited duration. In some foreign countries, notably in America and on the continent, inclined planes have had a more lasting popularity; only in 1969 did the enormous Ronquières inclined plane on the Charleroi–Brussels Canal come into use, on which 1,350 ton barges can travel.

But looking back once more to the era of the small canals, one can consider again, just how much did the American Robert Fulton contribute? We can say for certain that fitting wheels to boats to enable them to travel over inclined planes was his own idea, as also was his design for the bucket-in-the-well method of lifting. And it was not really his fault that the latter was not fully successful. Thus it can be said that the Bude Canal owed much to the ingenuity of an eighteenth-century American. It must, however, be also said that the development of canal inclined planes owes as much, if not more to the Salopian William Reynolds, to Edmund Leach, to the Earls Stanhope and to James Green.

BIBLIOGRAPHY AND REFERENCES

Dickinson, H. W. *Robert Fulton, Engineer and Artist* (1912).

Fulton, Robert. *Treatise on the improvement of Canal Navigation* (1796).

Hadfield, Charles. *The Canals of South Wales and the Border* (2nd ed, 1967).

Hadfield, Charles. *The Canals of South West England* (1967).

Hadfield, Charles. *The Canals of the West Midlands* (2nd ed, 1969).

Hadfield, Charles and Biddle, Gordon. *The Canals of North West England* (1970).

McCutcheon, W. A. *The Canals of the North of Ireland* (1965).

Smith, Peter, ed. *Waterways Heritage*, pub Luton Museum and Art Gallery, 1971, page 49. (Chard Canal inclined plane).

Tew, D. H. 'Canal Lifts and Inclines', *Transactions of the Newcomen Society*, XXVIII, 1951–2 & 1952–3.

Williams, W. Howard. 'The Canal Inclined Planes of East Shropshire', *Journal of Industrial Archaeology*, 2, No 3, Oct 1965.

Index

References to illustrations are printed in italics